# So You Want to Be a Scientist?

# So You Want to Be a Scientist?

Philip A. Schwartzkroin, PhD
*University of California*
*Davis, CA*

OXFORD
UNIVERSITY PRESS

2009

# OXFORD
UNIVERSITY PRESS

Oxford University Press, Inc., publishes works that further
Oxford University's objective of excellence
in research, scholarship, and education.

Oxford   New York
Auckland   Cape Town   Dar es Salaam   Hong Kong   Karachi
Kuala Lumpur   Madrid   Melbourne   Mexico City   Nairobi
New Delhi   Shanghai   Taipei   Toronto

With offices in
Argentina   Austria   Brazil   Chile   Czech Republic   France   Greece
Guatemala   Hungary   Italy   Japan   Poland   Portugal   Singapore
South Korea   Switzerland   Thailand   Turkey   Ukraine   Vietnam

Published by Oxford University Press, Inc.
198 Madison Avenue, New York, New York 10016
www.oup.com

Oxford is a registered trademark of Oxford University Press

Library of Congress Cataloging-in-Publication Data

Schwartzkroin, P. A. (Philip A.)
  So you want to be a scientist? / Philip A. Schwartzkroin.
     p. ; cm.
  Includes index.
  ISBN: 978-0-19-533354-1 (alk. paper)

  1.  Research. 2.  Scientists—Vocational guidance.  I. Title.
  [DNLM: 1.  Research. 2.  Career Choice. 3.  Science.  W 20.5 S3995s 2009]
  Q180.A1.S338 2009
  502.3—dc22
                                    2009000942


9 8 7 6 5 4 3 2 1

Printed in the United States of America
on acid-free paper

*To my wife, Gerry Chase, whose support made this book possible—and whose critical eye and ear have made it much better than it would otherwise have been.*

# PREFACE ◆

This slim volume is not intended to be a "how to" book. Rather, it consists of a set of reflections, formed over a career of 30+ years in the laboratory, on various aspects of the *job* of being a scientist. In that sense, it is a highly personal response to the challenges and opportunities of research science. I hope that it also deals with a number of issues that most, if not all, scientists-in-the-making must confront.

I should say something at the outset about what I mean when I say "scientist." Specifically, I'm referring to those who do research for a living. Various methods of research have been discussed at length by others who have far more philosophical training than I do. This book is not about scientific method or research protocols: it is about the job of doing research science.

Why write a book like this one when there are so many high-quality training programs that are devoted to turning out top-notch scientists? The answer to this question rests in my belief that most training programs, while emphasizing how to do good research, neglect the fact that "being a scientist" is indeed a job—and that there are many facets of the job beyond simply doing good research. I know as a certainty that many of the issues I address in this book were not explicitly taught when I was in training. Rather, the expectation was—and I believe still is—that young scientists learn the business by observation and mimicry. Despite all the books and journal articles that explain the how-to bases for scientists-in-training, becoming a scientist is (and always has been) basically an *apprenticeship* process. Many of the most important things you learn during training are not written down anywhere. You learn them, and about them, as you watch your teachers and mentors navigate through the system. And many of the key insights come as a surprise to the naïve and well-meaning student who somehow believes that "doing research" is an occupation that is substantially different from doing a job in "the real world."

These surprising revelations often appear only after the trainee—and the training institution and mentor—have devoted a significant amount of time, effort, and money to the educational/training process. While some students recognize and accept these important ancillary aspects of the profession (and become very adept at dealing with the non-scientific demands of a laboratory career), others discover that the "business" is not what they expected to be signing up for—and either quit or become unhappy and unproductive. Because many students commit to this process without adequate knowledge, the decision to pursue a career in science—for both the student and the training environment—involves a substantially high-risk investment.

One important general message from this book is that being a scientist is very much like other jobs. Like it or not, scientists are very much a part of the real world.

What I offer here are some observations and commentary—and yes, in some cases, advice—that are aimed at making young scientists aware of what's coming. This book is meant for those who are considering a research career or are early in their training. My starting point is the conviction that the more you know about the process, the lower the risk you take in entering the field and pursuing this career. The more you know, the better the odds that you will make decisions that are consistent with your long-term goals. The more you know, the more likely that you will be successful and happy.

As noted above, this book is based on my personal experience. My own research career has been entirely within the academic biomedical community, and so many of my observations are colored by that environment. I make no pretense of presenting a complete and/or unbiased picture of the research field. But I believe that the issues I've encountered in my own area of work are relevant to the general scientific community. We all are engaged not only in doing experiments, but also in writing papers and grant applications, in giving talks and training students, in gaining promotions, and in making a significant impact in our chosen area of study.

We are all, also, a part of the greater social fabric, and so need to be concerned about the place of science in our society. Not that many years ago, as I was entering the field, the title "researcher" or "scientist" was held in high esteem. I fear that that esteem has been considerably eroded over the past couple of decades. Why that should be, and what we as scientists can do about it, also deserves some discussion. While I don't pretend to provide any answers to those questions in this volume, I do hope to spark some thought and discussion, particularly among the young people whose job it will be to restore the research scientist to a position of respect and leadership.

# ACKNOWLEDGMENTS ◆

As you will undoubtedly see as you read through this book, I believe strongly that science is a social enterprise, and that the social context in which we do our research is critical to our success. It is fitting, therefore, that as I start this book, I offer my gratitude to the individuals who have shaped my career and guided my research endeavors. My thanks to the many teachers and mentors—John Fuller, Richard Wimer, Charles Gross, K-L Chow, and Per Anderson, among others—who stimulated my interests, encouraged me, and supported my efforts. In particular, I owe a huge debt to David A. Prince, in whose laboratory my research career first realized its current form. My thanks also to the many scientific colleagues who form a supportive and challenging network within which to think and explore. I would like to offer special thanks to my scientific partner and friend of 15+ years, H. Jurgen Wenzel—a true scientist's scientist. Finally, but most importantly,

I'd like to recognize the many students and fellows who have gone through my laboratory. I've learned a tremendous amount from them—not only about teaching and mentoring, but also about how to be a scientist. It is from them, and for them, that this book took shape.

# CONTENTS ◆

# So You Want to Be a Scientist?

# 1 ◆
# Getting Started

◆ General Comments—What Characterizes a Scientist?

Let's say you've begun to consider the idea of research science as a career but aren't quite sure that you're the right type of person for this job. Are there particular personal characteristics that lend themselves to this pursuit? My experience is that although the research field is populated by virtually every sort of personality imaginable, there are a few features that are obvious within the entire scientist population.

First, the people who have been successful as scientists are all curious about how things work—not *why* they work the way they do (that is, really, a religious/philosophical question), but *how*.

If you need to know *why*, don't become a scientist.

Science is the business of asking—and understanding, and explaining—*how*. The *why* and the *how* questions are not necessarily in conflict or opposition, but they are very different. "Scientific research" is a system for gathering information and interpreting that information to answer the *how* questions, but it has virtually nothing to say about *why*. That is not to say that scientists, as individuals, cannot (and do not) have personal views on *why*, but investigating and defending those views are not part of their scientific function.

A second general feature of successful scientists is that they are imaginative, often in a puzzle-solving way. Their training teaches them to think within a given system, but they like to push the envelope and see how far they can stretch it. At the same time, these individuals tend to be highly disciplined, a characteristic that helps them work effectively within the chosen scientific system. That discipline tends to include careful attention to detail, a doggedness in pursuing questions of interest, and a significant tolerance for failure and frustration—since most experiments don't "work" exactly as initially planned, often for reasons not immediately detectable.

Inevitably, one of the most noticeable features of successful scientists is their enthusiasm for the question at hand. When you listen to successful scientists speak about their work, their excitement tends to be contagious. You get a sense not only of the intrigue and beauty of the topic, but also of the social relevance of work in the area. It seems likely that a major factor in the decision of such people to become scientists was their love of the subject matter.

Indeed, it seems to me that that feeling should be the determining factor for all of us who choose this career. As I hope will become clear in the chapters to follow, a career as a research scientist is not easy. The rewards, while significant, won't make you rich and only rarely will make you famous. If your major career goal is to make lots of money, there are undoubtedly better and easier ways to do it. For a happy career in research, you need to love what you do. And for those who enjoy dealing with conceptual or technical challenges and puzzles, research is rich in its opportunities.

So, then, how do you determine whether such a career is right for you? Of course, it helps if you're good at "it"—at least at some aspect of your field of interest. Your strength may be in a particular intellectual capability (really good at math?) or it may be a technical facility (outstanding hand–eye coordination?). It may be a unique insight into behavior, or it may be an ability to visualize complex structures in three dimensions. There are all sorts of special abilities that people have that might help them achieve success in research science. Recognizing such abilities in yourself, and choosing a field in which you can maximize their applications, may help minimize the inevitable doubts and frustrations associated with choosing your career.

How might one become aware that he or she does, indeed, have the "right stuff" for a scientific career? While some of us "know" from an early age what we want to do as an adult, most of us need to experiment and consider various possibilities. For my own money, I'd start with a broad and diverse education. Although at some point in your education, you will certainly need to specialize, a broad general education not only helps you discover your loves and abilities, but also helps you develop as a person—and every scientist should be, first and foremost, a person. What I'm saying here is that you don't need to be "pre-med" to choose a career in biomedical science, and you don't have to be a math major to pursue a career in astrophysics. What you do need is to experience a broad enough range of disciplines to know what you love and what you're good at.

◆ Preparation—Education and Experience

While one can take many different pathways toward a research science career, there are some common "requirements" without which the process is much more difficult than it need be. It's all fine and good for a young boy or girl to dream about the great discoveries he or she will make, but romantic fantasies are not enough. The career in science entails a significant number of years in preparation. That preparation normally involves two major components: formal higher education (i.e., college plus graduate school [or medical school]) and practical (laboratory) experience. Almost certainly, to establish yourself as an "independent investigator"—someone who makes his or her own decisions about what to study and how to study it—you'll need a Ph.D. or equivalent. While not all research positions require that level of preparation, there's no question that a good basic education in science is a key to research positions of responsibility.

From a practical point of view, experience is as important as this educational "entry ticket"—experience in the laboratory, in the field, or wherever your chosen area of research takes place. Early experience is part of the apprenticeship process. It not only serves a didactic purpose but also provides a taste of the work. It would be, after all, rather foolish to devote years of time and money and effort working toward a career goal that you knew nothing about when you started the process. I'm strongly in favor of early experience—in high school if possible, but certainly in college—well before you need to make a decision about what you want to do and how you want to do it. This early experience does not mean early specialization. Rather, your experience in high school and college should prepare you to think, to write, to work effectively with others. And it will provide a "safety net" that will help you move forward if you discover that a scientific career is not for you.

Clearly, the ideal time to experiment and think about various career alternatives is during undergraduate college days. Although there is considerable pressure now to declare a major early and to specialize as soon as possible, I disagree with that philosophy. I don't think it makes better scientists. And I do think that it robs the student of much of the fun and exploration offered in college.

"Crunch time" comes when you apply to graduate school, for that application process almost certainly means making at least a general decision about what you want to study. How do you choose *the* program that is right for you? It is important to realize that the choices made at this time are not irrevocable.

You can change your mind.

If you discover that you've made a mistake, you can change direction. So the choice of a graduate program may not be critical. But it *is* important, for this choice will largely determine the kind of exposure you get to your chosen field. It will certainly influence the kind of feel—and feelings— you develop for your work. And like it or not, the reputation of the program—not so much the school per se, but the specific graduate *program*—will form the basis of how you're initially perceived by others working in your field.

So how should you pick a graduate program? You would start, presumably, by finding a set of programs that reflect your interests. Does the school offer a degree-granting program specifically in your area of interest (e.g., geology, neuroscience, astrophysics)? Perhaps the degree-granting program is more general (e.g., biological sciences, physical sciences), but there is a strong core faculty in your area of interest. With that background information, you can then investigate the programs that appear most interesting from a personal perspective. For example, you might want to consider whether a large program or a small program is best for you. A large program generally provides broad exposure and many resources. It also offers a built-in set of colleagues who share interests and enthusiasms. But one can easily "get lost" in the crowd and not receive the desired personal attention. In contrast, smaller programs are usually more limited in the opportunities they offer. However, they may provide more frequent personal interaction with senior faculty and enable students to focus on a more restricted area of investigation. There is no one "right" program: each program offers a different set of advantages and disadvantages, so it's important to explore the options.

Good choices depend on information.

Study the program descriptions, visit the schools if possible, see what feels right. But take note: all programs treat graduate students as adults, so be prepared to take responsibility for your choices.

In most graduate programs, you'll have some time to "get acquainted"—that is, take introductory courses, meet some of the faculty, learn more about who's doing what kind of research. Typically, that introductory time includes sampling the activities of several different laboratories (i.e., short lab rotations). And often, by the end of your first year, you'll be asked to choose a home laboratory and advisor. Again, those choices are not irrevocable, but they are important. The choice of advisor—someone who will act as your mentor through your graduate studies—can have long-term consequences. The advisor's laboratory will determine the context for your research activities. The head of the laboratory provides a powerful model—positive or negative—with respect not only to research focus but also to research "style" (yes, every scientist has a personal style). So when you identify a laboratory, you should be thinking not only of what type of research you might do there, but also whether your personal style fits with the style of the laboratory. A good fit can mean you'll enjoy yourself. A bad fit may mean that you spend the subsequent 4 or 5 years wishing you were elsewhere—and may even have the effect of "turning you off" with respect to a future as a scientist.

In general, your experience as a graduate student will set the tone for your subsequent work. And your advisor—if truly a supportive and sympathetic mentor—can be a powerful ally in your introduction into the field. His or her connections provide the basis for your initial connections. His or her reputation will affect how you are perceived as you emerge from your graduate work. You may switch areas of interest after graduate school, but the influence of your graduate mentor will mark your scientific style and thinking for years to come.

Once you settle into a laboratory, you'll inevitably begin the agonizing process of determining the focus of your dissertation research. Usually, the outlines of a dissertation project emerge gradually, influenced by your own interests, the focus and techniques of the lab in which you're working, and a number of indeterminate (and sometimes surprising) factors that, for whatever reasons, grab your attention and that of your advisor. Usually, dissertation topics are "conservative"—solid, technically accessible (for that laboratory) pieces of work that are achievable in a defined period of time (i.e., not open-ended). The work should be intellectually interesting (after all, you'll spend at least 2 or 3 years working on that problem) and sufficiently challenging so that you learn many of the technical approaches commonly used in your field. A dissertation project gives the student a glimpse into a given field, provides a context for the student to learn experimental techniques, and serves as a platform for further study. In some cases, especially when the topic is slightly out of the area of expertise of the mentor's lab, a dissertation project gives the trainee an opportunity to develop a special research niche, separate from that of the adviser.

While dissertation research projects are often scientifically significant (and certainly important for initiating the trainee into the field), Ph.D. dissertations rarely establish a young scientist's reputation. Further, your dissertation project does not relegate you to a lifetime of work on that topic. Indeed, more often than not, the student will choose a direction for subsequent research that differs substantially from the topic of the graduate dissertation. Indeed, I did my dissertation research on the neurobiology of the visual system but have spent my career studying the cellular bases of seizures. So I don't think that it's worth worrying too much about if or how the field will "brand" you on the basis of the theme of your graduate research. What is of concern is that you do a good job—that is, publish at least one high-quality research paper on the basis of your studies, and gain respect from those working in the area of your research.

Trainees often have trouble finishing their dissertations. They want to produce something special, and that drive is reinforced by the trainee's mentor, who wants his or her laboratory to publish a highly cited paper. Publications are a real and important part of your credentials, even as a student. It's the one way of making an impact on the field. So it's natural for the student to feel the effects of two competing pressures: on the one hand, the draw toward publishing quickly and finishing; and on the other hand, the desire for adding additional data to improve the study. As you near the end of your dissertation research, it may be helpful to recognize that no experiment of any consequence has an obvious end-point: there is always more to do. It's therefore important to pick a reasonable stopping point (perhaps arbitrarily), finish your dissertation research, write your paper(s), and "get out."

All of your graduate training points you toward this goal of "completion." You will attend seminars and conferences and meet the people who are working in related areas of research. You will present talks and posters. You will, basically, *practice* the art of being a scientist.

And when you've learned the techniques you need and have the confidence that's so important for independent work, it's time to get your degree and move on. The Ph.D. (or comparable degree) is your membership card that gets you into the society of scientists. It puts you in a position to start exercising intellectual choice, to take responsibility for research design, and to receive credit for significant research contributions. It gives you—perhaps magically—the critical start that we all need toward independence.

What does a Ph.D. actually mean? Other than serving as a magic pass, what does it symbolize? Every mentor, and every student, probably has a different set of answers to that question. My answer—the one I've always used to determine when one of my students was ready to move on—is based on the view that a Ph.D. means scientific/research competence. As such, the Ph.D. says to the world that this individual is intellectually and technically capable of independent research. There are other contributors to this end state—that the trainee has published a significant piece of work, that the lab has little more to teach him or her (at least technically), and so forth. But the bottom line really reflects the "fact" that the trainee is ready to strike out on his or her own.

◆ The Next Step—Postdoctoral Training

Your real career as a scientist starts *after* you receive your Ph.D. Until then, you are "just" a student. But with the Ph.D. in hand, suddenly you are functioning (or so the belief goes) at a professional level. In that sense, the Ph.D. degree is truly magical, allowing the former student to enter a land of great opportunity. From my perspective, the postdoctoral period can be the best time of one's scientific career. Classroom demands are behind you, the time pressure of finishing a dissertation is a thing of the past, and you can now focus on what is presumably your real love—research in the laboratory. But the postdoctoral period also carries some heavy expectations and consequences. Unlike your earlier research, postdoctoral output will identify you with a particular area of investigation, and with particular research tools. As a reflection of your first real "independent" work, publications and presentations from the postdoctoral period also serve as your more formal introduction into a specific research area, giving others in that field a sense of your direction, professionalism, and passion. Much more than your graduate work, postdoctoral output sets the tone for your future. Given this reality, choosing postdoctoral positions (and there may well be more than one) is obviously important. As you make this choice, there are several difficult questions to consider (although I'm not sure there are correct answers to any of them):

1. *Should I stay at the same institution (and perhaps with the same advisor) where I did my graduate work?* The "pros" may include personal considerations and attachment to the geographic area, comfort with the lab and relationship with the mentor, opportunities for collaboration, and involvement in an exciting area of research. The "cons" include giving up the potential advantages of new relationships, missing out on the excitement of exploring new questions and taking on new challenges, and forgoing (or at least delaying) the technical advantages that might be offered by a new laboratory. Staying in the same place may complicate (or simplify) your decision about continuing to work along the same lines of research—or, indeed, make

that decision completely moot. Most laboratory mentors discourage their Ph.D. students from staying on after they've completed their Ph.D. studies and encourage them to seek new experiences and positions, but are often welcoming if the fellow returns after some time in another laboratory or institution.

2. *If the decision is to look elsewhere, should I stay within this country or seek a position abroad?* The attractions of working abroad are considerable, including exposure to new cultures, new people, and new ways of thinking about scientific problems. The physical separation may, however, make it more difficult to stay connected with a given research area in one's home country, and therefore make eventual job searches more challenging.

3. *Should I focus on learning new techniques and branch out into new lines of investigation, or should I try to refine the techniques I know and deepen the involvement related to my previous research experience?* To be sure, especially as a young scientist, breadth of interests and of research capabilities is not only personally exciting but also potentially attractive on the job market. However, as our scientific society becomes more and more specialized, it is more and more important to do at least one thing very well—and to be identified with that approach (or problem or technique).

4. *Should I look for a postdoctoral mentor with seniority and experience, or should I "hitch my wagon" to a young, up-and-coming investigator?* Senior investigators often offer positions in larger labs, with multiple projects. They have track records of mentoring young scientists, which you can check out by talking to previous trainees who have emerged from that laboratory. And these senior scientists often have broad, international reputations (and contacts), which can be quite useful in helping the young investigator establish himself or herself in the field. On the down side, however, it is easy to get lost in large laboratories, and to be viewed as simply a set of hands for the lab leader. Working with a younger

investigator may offer more opportunities for independent thought and work, as well as more intimate relationships with the lab leader and other lab personnel. However, financial support is usually more limited; for example, the principal investigator of a small lab is less likely to provide access to the most current (and often expensive) facilities.

5. *How much does the institution matter?* When assessing your training, outsiders will generally look more closely at the laboratory affiliation than at the institutional affiliation in forming first impressions. However, in terms of later job search efforts, it is sometimes an advantage to work within a large institution that potentially could offer you a faculty-level position when you're ready to make that move. Of course, the desirability of such opportunities will depend on the kind of postdoctoral experience you've had at that institution.

However you answer these questions, a key goal for your postdoctoral research will be to carve out a niche in the field that you've identified for your future research efforts. Your lab work, and the publications and presentations that are based on that work, will define you—opening (and closing) doors of opportunity. It is this work that differentiates you from all your generational colleagues (and competitors). Granting agencies, potential employers, and senior investigators in your field will all evaluate your applications and inquiries based on this association. Postdoctoral work will also give you an opportunity to become *the* expert in your chosen niche area of research. Perhaps most importantly, developing such an area of expertise will give you the opportunity to delve deeply into a question of significance, to develop your investigational skills, and to find the excitement and rewards that research in novel areas of investigation can offer.

How does a young, beginning scientist choose such a topic? How can you tell what research direction is likely to bear fruit over the long haul? For guidance and assistance, you'll inevitably depend on a mentor—usually your post-doctoral advisor—who has more experience and a broader view of the field than you do.

> Choose your advisor carefully, and cultivate
> the relationship.

For this reason alone, the interaction with your advisor/mentor may well be the most important professional relationship you ever have. Work toward establishing a close and interactive relationship, marked by professionalism, mutual respect, and appreciation for the insights and advantages your mentor can offer you. As I indicated above, your mentor can—if he or she appreciates your talents—open many doors. He or she can introduce you to "powerful" people, advocate for you in responses to reviewers and granting officials, and provide the all-important exceptional letters of recommendation. Most importantly, he or she can help you negotiate the labyrinth of experimental choices. His or her model may well be the one you find yourself following as you start your own lab. Recognize, too, that your relationship with your mentor may well experience "hard times," as do many close, long-lasting relationships. Try to avoid the distractions and interference of personal differences, and focus on the pluses. Be patient with him or her, and realize that as is often also the case with parents, advisors appear to get smarter and smarter as you (the trainee) mature and take on responsibilities that you didn't even realize existed when you were "young."

Of course, sometimes the mentor/trainee relationship just doesn't work. It is therefore important also to cultivate other associations and contacts. Science—even for the "lone ranger" in the laboratory—is basically a social enterprise: the more people you know, the more support you can muster and the more effective you can be as a researcher.

It is perhaps an obvious fact that relationships with key figures in your field can significantly facilitate your career. It's perhaps not so obvious, but equally true, that relationships with peers and even younger colleagues can and will open new research doors and make your research life more enjoyable. It's easy to see research as an activity characterized by facts and numbers and impersonal research tools. But the fact is that your research—not only the problems and the data, but also the quality of the experience—will be substantially affected by your human interactions.

There are many ways to facilitate collegial interactions, and they're all integral parts of a research career. Go to meetings and meet people. Make presentations at seminars and interact with other participants. Publish and interact with reviewers and editors. Submit grant and fellowship applications, and get to know the people who administer these programs. These relationships will constitute the network by which you can market yourself.

> "Self-marketing" is, for better or worse, a major
> part of the job of being a young scientist.

Start developing a resume—your curriculum vitae (CV)—at an early stage, and do your best to generate a document that reflects you and your accomplishments. By all means, make it impressive. It is a sales document, to be used in presenting yourself and your credentials to prospective employers, potential grant/fellowship review committees, and coworkers and colleagues. But keep it focused and relevant. Most of those to whom you present your CV will have lots of experience in reading between the lines, so padding (length for the sake of length) is usually not very helpful. Don't risk losing the reader's attention by presenting too much information. And remember, you can "help" the reader focus on especially impressive accomplishments by organization and emphasis.

◆ What Does the Young Scientist Need to Learn?

Throughout the period that I've described above, the young scientist is basically "in training." The graduate and postdoctoral periods constitute an extended apprenticeship. Given that the period of formal education and apprenticeship is long, it is perhaps reasonable to ask what one needs to learn. For some reason, this question never seems to be made explicit in most training programs. Although the specific information that is "learned" during training varies widely from field to field, there are a series of "practices" that can be generalized—and the following chapters will concentrate on those practices. A brief list will provide some sense of the goals of training:

- To learn to *think* as a scientist does, including such important practices as:
  - Formulating appropriate (i.e., experimentally addressable) questions and hypotheses
  - Designing experiments to test hypotheses
  - Differentiating important questions/topics from busywork
  - Critiquing work (yours and that of others) sympathetically but incisively
- To establish a laboratory and to run that laboratory according to the rules of responsible scientific conduct
- To communicate with the scientific community and with others who support your work in a variety of ways, including:
  - Presenting an interesting and informative seminar or conference talk
  - Conveying data effectively in posters and abstracts
  - Writing papers that are viewed as important contributions to the field
  - Constructing grant proposals that speak to important issues and are likely to be funded

- To interact effectively with mentors, peers, students/trainees, and other colleagues
- To conduct yourself as a "good citizen" within the scientific community

Contrary to popular perception, these goals of training are all complex and difficult. Indeed, when we step back and consider the list of activities in which one needs to gain expertise, it is not at all surprising that the training period is so long. All require time and practice, and all can be mastered most efficiently with the help of a model (presumably your mentor). No matter how "smart" you are and how much information you've absorbed on a given topic, actually applying what you know to solve problems and learn more about our world is not an intuitive process. So it is important to be aware of these aspects of a research career when you're considering a life as a scientist.

◆ Are There Special "Secrets" to Success?

The chapters that follow take up many of the goals identified in the paragraphs above, but there are a few "tricks of the trade" that are worth noting at the very start. The following list is certainly not exhaustive, and it reflects my personal experience. But in talking with my students and fellows, I often find myself repeating these same key points:

- *Maintain your focus.* Especially when just starting in a scientific career, choose a limited set of research questions and maintain your focus on those questions. Most of us who choose science as a career are excited by many of the issues that can be addressed in the laboratory, and it is all too easy to find oneself involved in multiple research projects, spread out in several different research directions. However, except for a few extraordinary individuals, few of us are capable of generating top-quality work simultaneously in many different areas of endeavor. Maintaining focus—both intellectual and technical—is difficult but extremely useful as one builds a career and reputation for expertise. Most mentors will work hard to keep young trainees "on track," and there's a reason for doing so.

- *Take a broad view* of the area in which your research is embedded. Perspective tends to be something attained gradually. It's difficult, as a young investigator, to master even a small part of the puzzle. But scientific research is an activity that involves not only describing each piece, but also fitting the little pieces together to form an interesting and significant "whole." It's essential to have a sense of where your little piece actually fits. Fitting pieces together is part of the cooperative nature of research science. The idea that there are "big-picture" thinkers and "detail-oriented" thinkers has some validity, but a successful scientist needs to do both.

- *Read the literature.* One way of generating a good "big picture" is by reading the literature. This suggestion may seem obvious, but given the vast (and rapidly growing) literature associated with most research topics, this job is increasingly daunting. It's nonetheless important to keep abreast of recent papers in your field and to learn something of the research history (i.e., to get a sense of how we've come to our current state of knowledge). Familiarity with the literature not only helps you to develop that big picture but also helps you to avoid a major pitfall in research—"reinventing the wheel." There is some truth in the view that "there's nothing new under the sun," but you can certainly try to put a new face on it, gain a new perspective, add new details. You can't do that without reading about prior work in your area (or related areas). Finally, if you don't know the literature, you run a significant risk of insulting other (previous) workers who might have already "discovered" something you want to claim as your own. Familiarity with the literature gives you a means of evaluating the work of coworkers and colleagues and serves as a major gateway to the ongoing communication that characterizes every aspect of research.

- *Communicate!* While the emphasis in research science is generally on actually doing the research, effective communication is key. Given the collaborative and interactive nature of research, communication with your research colleagues (e.g., mentors, students, peers, collaborators, technicians) is critical for a successful research career. Outstanding research accomplishments may be wasted if you don't communicate them effectively to others. More to the point, outstanding research accomplishments may be difficult to realize if you are not "in touch." This practice takes many forms (e.g., papers, talks, phone calls, e-mails) and can be done to varying degrees, but there's no question that advances in modern research depend on effective communication. Remember, too, that your reputation depends on others being aware of what you've done and are doing.

- *Respect the hierarchy.* Innovation and drive are generally encouraged in young scientists, and we do tend to reward those who exhibit ambition and self-confidence. However, as indicated above, research training is an apprenticeship process, and the entire research structure is based on a not-too-subtle hierarchy. The insights and wisdom of those more experienced in the field deserves respect, even though young people invariably think they know more than their elders (that seems to be true in all walks of life). Whether that's the case or not, the viewpoints of more senior investigators can be valuable. In training young scientists, senior scientists generally encourage (indeed, train) their students to argue in support of their own ideas, experiments, and data interpretations. One would hope that the trainee is able to distinguish between being respectful and being in unthinking agreement. You don't have to agree with the views of your mentor (or other senior researchers), but you should consider them carefully and respectfully. If, in that context, you can persuade them of the advantages of your own viewpoints, you will have gained their respect, and that respect will be an essential part of your reputation.

- *Network.* An increasingly important part of the communication process is networking. "Networking" has become a buzzword in our society, and for good reason: almost every aspect of our lives has become so complicated that no one person is able to master every aspect of a given issue. Nowhere is that more true than in addressing research questions in the laboratory. Scientific research is now, almost by definition, a collaborative process. Gone are the days (if they ever really existed) when the scientist worked alone, and by power of his or her individual imagination and brilliance produced major advances in our knowledge and understanding. Today, very few individuals can master all the information and techniques required for an effective assault on a research problem. So it is important to establish connections and to share ideas, results, and even data. In the "old days," young investigators were often cautioned against such sharing, for fear that the laboratory would be "scooped"—that is, that someone else would steal the idea and publish the critical experiment before the lab of origin had an opportunity to do so. That is, admittedly, a danger. But it seems to me that in the current scientific environment, the advantages of sharing far outweigh the risks of being scooped. Besides, the downside of being scooped is far less damaging than a reputation as a selfish (even paranoid) investigator.

You'll note that in virtually all that I've written above about tricks of the trade, there's been an emphasis—explicit or implicit—on how your actions affect the development of your reputation. I will, throughout this book, make that point over and over again.

> Your reputation, whether as a scientist, as a teacher, or as a colleague, will always be your most important professional asset.

Consider what you'd like that reputation to be. Don't assume that simply by doing good research, you've taken care of it.

---

**Real-Life Problem**: You're a college junior and thinking about a research career in biomedical sciences. You like your biology classes, but all your practical research experience has been in a human psychology laboratory. You've got a job lined up for the summer in the same psychology lab that you've worked in for the past year, but you're wondering about applying to graduate programs in molecular biology. How should you spend your summer?

1. Forget about the job in the psychology lab, and seek a position (even as a volunteer) in a molecular biology lab.
2. Proceed with your plans, on the assumption that a significant amount of work in a given lab—perhaps resulting in a substantive publication—will help your application to graduate school, even in a different subject area.

---

3. Take some time to clear your head. Be clear about what you like—or think you like—about molecular biology versus psychology. Try to imagine what it would be like to spend your life doing work in either type of laboratory.
4. Talk with your professors and with graduate students in both subject areas.

**Discussion:** For this and the Real-Life Problems in the remaining chapters, there is no one "right" answer. Rather, many of the listed possibilities (as well as other alternatives) may be useful—and some of the choices are not mutually exclusive. For example, in the present example, both #3 and #4 are very important, whatever your choices. It is important to be as clear as possible about what you want (or think you want) and to talk with others who may help you to see which choices will best meet your personal goals. In the case of choice #1, there may be some practical issues to consider. For example, do you need to make money from your summer job? Are there relevant positions available to you? Depending on those considerations, #1 may simply not be a realistic alternative. Alternative #2, generating a significant piece of research work, is likely to have a positive impact on an admissions committee, even if the research is in a different field. However, most admissions committees look for some indication that the applicant has thought seriously about his or her choice of programs and has prepared appropriately. Further, it is important to

*Continued*

have some real sense—not just wishful thinking—of what you're getting yourself into. So if you're thinking about "switching fields," it would be useful to spend some time in an appropriate laboratory. You might even consider working as a lab tech in such a laboratory for a year or two after graduation before applying to graduate school. Then you'll know what potentially lies ahead, and you'll be well prepared to make your case to an admissions committee.

# 2 ◆

# Career Choices and Laboratory Nitty-Gritty

◆ What's the Right Position for You?

Much—most—of the information in the following pages is pretty general, and I hope it will be useful to a consideration of many types of research positions. To be sure, some of it is aimed particularly at those who are considering a research career in an academic setting—especially since most young scientists initially associate scientific research with academics and think of an academic position as their first career option. In addition, much of the content of the following chapters is particularly applicable to those who are considering the possibilities and responsibilities of laboratory leadership—that is, as a principal investigator (PI). But there are a variety of positions available to the research scientist, and as one progresses through various stages of training and preparation, it is important to consider the full range of possibilities.

So, then, what kind of position is right for you? Not everyone needs to be in charge of running a laboratory, with all the responsibilities associated with that position. Indeed, not everyone needs to earn an advanced/doctoral degree. In the paragraphs below, I'll consider some of the features of various research positions—not to make recommendations or try to influence decisions, but rather in an effort to raise awareness. I do so as a result of having seen many bright and capable students working reluctantly toward what everyone has assumed is the only worthwhile goal in the research enterprise—to become a laboratory PI. Many of these students have much to offer but really don't want the pressures of "PI-hood." And many of them, rather than finding a good match in the spectrum of research positions, simply give up research in frustration. Instead of potentially finding yourself in such a position, it's far better to gather sufficient information about your options at the start and then to find a comfortable niche that matches your abilities and personality. Successful laboratory research depends on contributions at many different levels, all of which are critical.

◆  The Principal Investigator (PI)

The PI is the "face" of the laboratory. He or she not only directs the research program but also is primarily responsible for selecting, developing, and implementing the research themes of the lab. In an academic setting, this responsibility is usually associated with an appointment in a professorial track and carries with it the assignment of laboratory space. In industry (i.e., for-profit companies), at research foundations, or even in government institutions, the position may carry a different title, but the responsibilities are strikingly similar. At least with respect to research issues, "the buck stops here"—that is, with the PI—across all these types of institutions.

In addition to identifying the research themes and goals of the laboratory, the PI is responsible for obtaining funds to support the laboratory effort. That may mean applying for research grants to outside agencies. It may mean developing a convincing research proposal for internal institutional consideration. It almost always means working out the research steps that are needed to reach the goals—general and specific—that justify the existence of the laboratory. Toward those research ends, the PI is the person who generally hires (and fires) the personnel needed to carry out the research program for which the laboratory is funded. This aspect of the work may also include, in the academic setting, training students and fellows in the intellectual and technical tools that are required for successful research. Finally, a no-longer-insignificant aspect of the job involves the verification that the laboratory complies with all regulatory issues involved in the research.

These responsibilities require the PI to develop a critical set of skills (not all of which are "scientific") and to exercise an important set of "personality traits." Aside from scientific acumen, then, what personal features should a good PI exhibit? Certainly, the PI needs to be a "take-charge individual." He or she needs to be able and willing to take on significant responsibilities and to be comfortable not only basking in the glory of success but also accepting the responsibility for failures. As a result, these individuals realistically must be willing to devote large amounts of time to their work. Running a laboratory is rarely a 9-to-5 type of job. Each experiment tends to have a life of its own, and the time commitment to the operation of the laboratory is considerable. Further, given the competitiveness of the field, the PI should ideally be able to accept criticism and to respond to it without taking negative comments personally. The PI—as all scientists—will need to be able to deal with frustration and have a boundless determination to see things through to a successful conclusion. And finally, the PI will need to balance personal ambition (all PIs are ambitious) with care and concern for colleagues, trainees, and employees.


◆   The Research Scientist

Let's say that you don't want the responsibilities of leadership and are not eager to run your own laboratory. That certainly doesn't mean that there is no place for you in research science. Indeed, the scientific world evolves only through the efforts of well-trained, hard-working scientists who actually implement the ideas and plans of the PI. These individuals have different names and different levels of responsibility in different settings (e.g., research associates, research technicians, project scientists); for the purposes of this discussion, I will call them research scientists. Who are these people? What do they do and how do they prepare for this job?

To start with, the research scientist is someone who enjoys science, but perhaps not as a 24/7 occupation. For many, the ability to spend a defined period of time on a job, without a total commitment—intellectual, emotional, —to that job, is an important consideration. Thus, even someone with many years of postgraduate education (including a Ph.D.) may choose to work for someone else, to employ those skills they've developed during their training in a setting that is familiar and rewarding—but also have a life outside the lab. Other research scientists may not have advanced degrees but enjoy the laboratory life—the technical challenges, the social interactions, as well as the conceptual energy. Some research scientists begin to work in laboratories when they are quite young and become extraordinarily expert in what they do. Indeed, a senior research scientist is often the most important person in a lab because of his or her technical know-how and understanding of how the nitty-gritty of the lab—and the "system" in which the lab is embedded—actually works. Depending on their background and experience, such individuals may take on considerable technical and conceptual responsibilities and are often the individuals who direct the day-to-day operations of a laboratory. Many senior PIs—this author included—depend significantly on the skills, insights, and experience of such individuals. Despite the full involvement and commitment of these scientists, however, the work of the laboratory is generally associated with the PI. Because the research scientist is not ultimately in charge of laboratory operations, he or she is rarely credited with the successes of a research program.

◆ Choosing a Path

What are the issues that you should consider in deciding how you'd like to spend your time and effort? What will be sufficiently fulfilling yet not overwhelming? It's unlikely that you'll be able to answer these questions until you have a little experience in a research setting and are able to observe individuals who play these various roles. But here are some issues to consider and keep in mind:

1. What is the level of your commitment to research? Are you willing to spend 16 hours a day, 6 days a week, developing your research program?
2. How ambitious are you? Do you really want to be a star? To be the boss? Do you want to be in a position that will enable you to obtain awards and honors?
3. How good are your people skills? Are you comfortable dealing with personnel issues—not only hiring, but also disciplining and firing?
4. To what extent do you want to mentor students and fellows? Are you a teacher? A good model?
5. What is your "vision"? Do you have the creativity to drive your own research agenda?

6. How good are your organizational capabilities? Can you deal with (and be responsive to) rules and regulations that dictate the way in which research laboratories operate? Can you meet deadlines? Can you deal with red tape? Are you OK with issues like financial disclosure and conflict-of-interest inquiries?
7. Are you a good multitasker? Every PI needs to have several irons in the fire, and keeping track of the various lines of activity can be challenging.
8. Are you a leader? Are you a good judge of people? Do you like to be in charge, or are you more comfortable in the background?

◆ Alternative Science/Research-Related Careers

For those who love science but not necessarily laboratory research, there are careers outside the laboratory that can be fulfilling and that certainly contribute to the science goals of our society. Indeed, these careers can be preludes to, or "second careers" after, a time in the laboratory. Those who follow these paths often have graduate degrees and may develop significant expertise in areas of interest. These professions, just as careers in the laboratory, involve significant personal investment and can't be developed successfully without considerable love for science and research. If you're unsure about a career in the laboratory, consider the following investments:

• Science teacher
• Science writer and/or illustrator
• Science reporter
• Science lobbyist
• Physician or other medical professional
• Journal editor
• Grants manager or administrator
• Foundation administrator

**Real-Life Problem**: You are ambitious and smart but a bit happy-go-lucky. You have a good head for scientific research, but you're not very well organized and don't perform well under pressure. You're in the midst of your graduate training and trying to decide if you will pursue an academic career in which you'd be a laboratory leader. What factors should be considered in determining your decision?

1. How unhappy would you be working for someone else and letting the "boss" deal with organizational issues and the pressures associated with running a lab?
2. How important is it for you to be recognized for your work?
3. Is it important for you to determine the focus of your research, or do you simply like to work in the laboratory whatever the subject matter?
4. Can you see yourself developing self-discipline and making a serious commitment to dealing with the nonscientific issues involved in running a laboratory?

**Discussion:** Each of these questions is important, and each should be explicitly considered as you move through your graduate career. It is likely that you won't need to make this type of decision about your career choice right away. Indeed, you might consider spending a year or two as a postdoctoral fellow, during which time you can gain a better sense of some of the responsibilities involved in running

*Continued*

your own show. My own experience suggests that those who want to be PIs will soon find themselves chafing to start an independent career. They will have ideas for their own experiments and look forward to a time when they don't have to depend on someone else for permission (or support) to initiate those experiments. They will find that working for a PI or advisor is restricting and at least sometimes annoying (e.g., when colleagues automatically associate work from your lab—perhaps your own work—with the PI). If you don't have these reactions to working for a senior PI, then you should seriously consider research pathways that don't require that you run your own laboratory.

# 3 ◆
# How to Think Like a Scientist

◆ Science as a Thought Process

Scientific research is based on a particular way of thinking, and that way of thinking sets it apart from other types of work. While clear thinking and lucid presentation of your thoughts are likely to be important features of whatever career or job you choose, "scientific thinking" forms the absolute basis for everything else you will do as a scientist—designing experiments, analyzing data, writing papers, and so forth. I do not believe that anyone is *born* with this pattern of thinking. Indeed, I believe firmly that it is a learned "art," something you may start to pick up (perhaps unconsciously) from your parents or teachers in grade school, but a habit that is subsequently learned and refined. If you are serious about considering a career as a scientist, you should work toward making this thinking pattern a conscious part of your educational program. As you study and learn, thinking like a scientist may become "automatic," but it should never be taken for granted.

So what is this process? Scientific thinking has several parts. First, there is scientific "logic." I use the term "logic" not strictly in the sense of the philosophical notion (although a philosophy course in logic is certainly good scientific preparation) but rather in terms of how one forms hypotheses, collects data, and interprets results. In general, in the laboratory we rarely deal with the deductive reasoning so commonly taught in the classroom. Rather, we tend to generate experiments to test a hypothesis, in the form of an "if—then" statement. In the best of cases, we strive toward determining the necessary and sufficient conditions for some statement to be experimentally supported.

But we virtually never—and I mean *never*—deal with statements of truth.

Scientists say things like, "The evidence supports Hypothesis X." We don't say, "We know X to be true." We leave that latter exercise to philosophers and theologians. As scientists, we like to set reasonably strict limits on what we accept as sufficient and necessary evidence to support a hypothesis. Therefore, we tend to subject our evidence to tests of statistical significance and present our results in terms of probability. Some people may think that we're simply trying to avoid giving a straight answer by using this process. But since scientists don't deal in terms of absolutes, it's hard for us to say anything without offering a set of qualifiers.

Second, there is a scientific method—or perhaps more appropriately, a set of methods. As mentioned above, we generally start our work by stating a hypothesis and then designing experiments to address the hypothesis. The hypothesis should be explicit and simple. Clarity and precision are important initially in stating the question, and then also in collecting the data.

> Indeed, often our biggest conceptual roadblock is uncertainty about what question we are trying to answer.

Sometimes, when the issue at hand is broadly defined, we end up doing experiments that might be described as "exploratory" rather than "hypothesis-based." In other words, sometimes we don't know enough about the topic to form a specific hypothesis and need to collect more data so that a hypothesis-directed experiment can be formulated. These exploratory studies are particularly important as novel methods are developed so that new types of data can be generated. An important recent example of exploratory research is the set of experiments carried out to describe the human genome. One would not characterize those studies as hypothesis-driven research, but rather as exploratory studies, to see what's there.

But even with such exploratory studies, it's useful to ask whether there are implicit general hypotheses underlying this research. For example, if we take the example of the human genome studies, one such underlying hypothesis is that the human genome is composed of DNA sequences that are similarly – or differentially - represented in all people.

A third aspect of scientific thinking revolves around how we treat data: what types of observations are acceptable as data? what assumptions and limitations are involved in our experimental procedures? how our experimental and data-collection methods determine the kinds of conclusions we can draw? For example, every scientist is taught that it's not possible to "prove the negative." We cannot prove the proposition that there are no pink ponies, even if we observe no such ponies in a sample of several million data points; after all, we might have missed one somewhere, or we may have been using the wrong tools or methods for collecting the data (e.g., our instruments may not have been sufficiently sensitive).

> Clearly, our methods of data collection determine
> the types of data we can generate.

And observations made at one time may be unacceptable at a later time, especially as our sophistication increases and our criteria for admissible evidence change. At the time we make our measurements, however, what we can say is that pink ponies are very rare (i.e., the probability of finding one is less than 1 in several million). Contrast this process with the relative conceptual ease by which we can disprove our hypothesis—needing only to find one pink pony to make that case.

A fourth and critical part of the scientific thought process involves maintaining a clear awareness of the difference between *correlation* and *causation*. Much of scientific research focuses on trying to understand underlying causes, and we are trained to be sensitive to the cause–effect relationship. In contrast, much of what nonscientists assume to be "causal" simply involves observations of correlation. A current example is the claim that vaccinations in early childhood are responsible for autism. While there is a correlation between the timing of early childhood vaccinations and the appearance of autistic behaviors in many children, this relationship does not constitute a cause–effect connection. Certainly, identification of correlation is sometimes a step in the process of determining causation. But to establish causation—with a stated probability—requires carrying out a set of carefully planned experiments that systematically eliminate other possible explanations for the correlated observations and show the necessary relationship between proposed cause and effect.

Confusion about the difference between correlation and causation is quite common in everyday thought.

Correlation forms the basis of superstitious behavior as well as the rationale for many assumptions and beliefs held by people not involved in the scientific process. When those beliefs involve issues of scientific interest (e.g., whether childhood vaccinations are the "cause" of autism), then it is our job as scientists to clarify the confusion and identify the type of studies that could provide insight into a potentially causal relationship.

It should be clear that "scientific reasoning" is a system. This process does not depend on hunches or intuition, although intuition may be very important in forming hypotheses and in designing experiments.

> Scientific thought deals only with ideas or proposed
> relationships that are *experimentally testable*.

For that reason, some issues are beyond the realm of science at a given time in history (although it is certainly possible that, as our methods for measuring and testing become more sophisticated, those issues may be legitimately approached with new scientific methods). In general, the scientific approach is deeply imbedded within the culture of a given society, so that the issues it deals with (the things it can and can't do, can and can't answer) are a direct reflection of the interests and capabilities of that society.

Scientific thinking involves a relatively structured and socially agreed-upon set of rules and processes, but it need not be a staid and automatic process. True, the rules can be taught. But when it is at its best, scientific thinking is an individual and highly creative process. No two scientists are likely to approach a given problem in the same way. The best scientists are able to make interesting and often unexpected connections among observations, thus generating hypotheses and experimental goals that reflect their imagination and insight. Scientific thinking involves harnessing one's imagination, not overriding it. How one approaches a problem and how one develops hypotheses and designs experiments evolve with new experimental data and increasing understanding of the problem. Experienced investigators often provide critical insights to this process, since they have the breadth of background that allows them to make connections across observations that may appear to be unconnected to a less experienced observer. And new investigators in a field bring different perspectives, backgrounds, and conceptual contexts to the problem area; they may see problems in a new and refreshing light that allows more revealing studies to be carried out.

In all cases, investigators inevitably integrate the obvious (and expected) data with serendipitous results and use their highly disciplined imaginations to create new hypotheses and ways of thinking about a problem. This combination of disciplined thought and creative imagination is the basis of our best science. This process (and its results) is heady stuff. It is not for the shy and the timid; it is not for those who are satisfied with the status quo or with the usual way of doing business. We can even say that science is an art form (see Chapter 9) that allows an individual to create new concepts within a system that calls for strict rules of interpretation and conclusion-making.

◆   The Challenge of Scientific Thinking

If, in fact, scientific thinking underlies all your work as a scientist, then how well you think (within this context) will in large part determine your reputation as a scientist. The scientific system presents the individual with not only a challenge but also an invitation to freedom and creativity. How you handle these challenges and exercise these freedoms will be at the very core of your success.

Therefore, I believe there are a number of "must do's" for young scientists to incorporate as they enter training.

- First, learn the rules—how to generate hypotheses, test them, and interpret the results.
- Second, be aware of the major issues in your field, and stay focused on working toward their solutions. Even as you focus your experimental attention on testable hypotheses, make sure you understand where those hypotheses fit into the "big picture."
- Third, at least in the beginning, exercise caution and care in dealing with experimental results. It's useful to establish an early reputation for conservative thinking. Such an approach at the beginning of one's career allows one to become more audacious as time goes by. If you're known to be careful and cautious, then the field will be more likely to take your more "risky" hypotheses seriously.
- Fourth, strive to establish a reputation as an incisive, critical thinker. Learn how to reduce complex issues and hypotheses into simpler components that can be experimentally tested. Learn how to recognize errors in the logic of other investigators (and yourself). Learn how to identify the salient, critical features of a given problem.
- Fifth, in all your critiques and comments about work by your colleagues, try to be constructive. While it's useful to point out errors, it's even more useful to offer ideas about how the errors might be corrected. Be tough (conservative) but fair. Generosity is a fault only if it allows mistakes to enter and pervade the field. And remember, we all have something to gain from the perspectives of our colleagues.
- Sixth, develop an appreciation for the aesthetics of scientific discovery. Novel ideas, well-done experiments, clever insights—these can all be quite beautiful. An appreciation of creativity and innovation in others will help you to become more creative and innovative yourself.

◆  Experimental Design and Interpretation

The most obvious and perhaps most important application
of scientific thinking revolves around the design of experiments and the interpretation of experimental results. As
indicated above, the experimental process normally starts
with the formation of some hypothesis. But even before
you start thinking about how to formulate a hypothesis,
it's worth spending some time considering why you want
to study a given problem or question. What is the rationale
that justifies your time and effort—and the resources that
you'll inevitably need to carry out your study? You should
be able to explain this justification clearly, not only to yourself but also to others. Are you addressing a problem that
is important to society (e.g., trying to develop a new treatment for a disease or a new source of energy)? Are you
planning to investigate something that has no immediate
application or consequence but may be important as a
building block for the future? Are your studies exploratory
or focused on a specific issue that demands immediate
attention? Have you chosen your research focus because it
is likely to be easily fundable? Are you doing a study just
because you think it might be fun, because you like to solve
problems, because the techniques are "cool?" I am not sug-
gesting that you should find a "right" answer to these
questions to justify your moving forward with a planned
experiment. However, you should be clear about and able
to express the basis of your investigative motivation, since
that clarity will not only assist you in formulating clear
hypotheses and experimental goals but also help you in
identifying appropriate funding for the experiments.
Answering these questions may also guide your selection
of topics for study in your laboratory.

Once you've clarified why you want to study a given
problem, it's time to formulate a hypothesis. A hypothesis
is basically a prediction—an educated guess (based on data

from various observations and past experiments, and your own insight into the problem) about the relationship between some experimentally defined variables, as a function of some experimentally applied manipulation. As previously indicated, for your hypothesis to be meaningful scientifically, it must be *testable* according to available scientific methods. That doesn't mean that other types of hypothesized relationships are not important or significant; it simply means that you, as a scientist, cannot draw any conclusions about the "correctness" (i.e., likelihood) of the hypothesis if it is not experimentally testable. All you can do with such non-testable hypotheses is offer a *personal* opinion (i.e., not based on experimentally tested hypotheses). Remember, too, that the results of experimental testing cannot determine the "truth" of a hypothesis, only the probability that your results have not occurred by chance.

Hypotheses need not be correct to be valuable.

Indeed, one could argue that if you already knew whether or not your hypothesis was correct, you wouldn't need to do the experiment. We do experiments to discover something new or to confirm (or not) some belief about which we're not sure. Good hypotheses are clearly stated, with an explicit identification of what relationships are to be tested. When generating a hypothesis, don't worry about whether the ensuing experiment will show it to be correct or not. Indeed, science traditionally has generated important insights by testing long-held beliefs that turn out to be just that—beliefs that are not supported by experimental evidence (i.e., the experimental data indicate that the hypothesis is "false"). Such experiments often give rise to argument and discussion that lead to additional hypotheses and experiments, and thus move the field forward.

Experimental design is a challenging topic. How you approach a particular experiment depends on many factors.

These include the techniques and expertise represented in your laboratory, concurrent work of other investigators studying the topic of interest, and available funds and resources. Despite these factors, and despite the wide range of experimental approaches that characterize different areas of research, a number of conceptual features are almost always key to appropriate experimental design. For example:

- Be sure that the results that your experiments will generate (i.e., your data) will actually address your hypothesis.
- Avoid the trap of assuming, within your experimental design, the validity of your hypothesis (circular reasoning).
- Be careful in your choice of experimental "subjects" (or samples), so that the study results are not influenced by a skewed or biased sample.
- Make sure that your study includes the necessary controls or comparison groups.
- Carry out studies that are technically feasible, not only for your own laboratory but also for other labs (so that your study can be replicated).
- Try to analyze your data in a blinded fashion—that is, the experimenter analyzing the data should be unaware of the source or sample group represented by any given data point.
- Collect sufficient data so that you can draw statistically sound conclusions (i.e., your experiment is adequately powered).

While there are, of course, many other issues to consider in experimental design, ignoring (or being unaware of) these particular points often makes it difficult to interpret the results of a study.

Experiments are designed to generate data. But results, in isolation, are generally meaningless. Even more important than data generation is the process of data interpretation. Often the challenge here is to identify, within an often-confusing set of results, which data are important—at least with respect to your hypothesis. In other words, you must determine which are the really salient results amidst all the "noise" that is typically generated by experimental manipulations. While there are several techniques for making that determination, the most straightforward approach is to *design* your experiment so that the results are clear-cut with respect to your hypothesis. Using appropriate statistical methods for uncovering significance is also almost always a critical part of "interpretation."

Once you've designed your experiment and collected your data, your job is to "interpret" your results within the context of your hypotheses—that is, do the data support your hypotheses or not? Or, asked another way, are your conclusions (your answers to the initially posed questions) supported by the data that have been generated by your experiment? In many cases, your results will not support your hypothesis (and indeed may show it to be false). What should you do, then, with your data—and your hypothesis? If your results do not support your hypothesis, you can then reject the hypothesis. If your hypothesis is appropriately stated, even such a "negative" result will provide important information.

> The fact that the data don't support
> your hypothesis is not a reason to
> reject your *data.*

Indeed, you may want to examine your data for new insights and directions that will help you formulate a new hypothesis. Exciting "serendipitous" discoveries are often a result of plumbing results that "don't fit" the hypothesis under study. Once you reject your initial hypothesis, you can begin the process of generating a new one. That process may entail modifying your old proposal—in light of your new data—and then subjecting your revised hypothesis to experimental testing.

If your data do support your initial hypothesis, that doesn't mean you've proved the hypothesis to be "true." As noted above, the scientific approach deals only with probabilities and likelihoods. And inevitably, a given experiment will deal with only a limited aspect of a complex issue. So even though you've supported your hypothesis, your work is rarely done. New experiments, testing and extending the relationships identified in your initial study, are now in order.

◆  Focused Investigation vs. the Big Picture

A well-designed experiment is usually focused and specific. It asks (and, if you are fortunate, it answers) a narrow question that can be "solved" in a way that invites relatively little controversy. In contrast, one *does* invite controversy, misunderstanding, and alternative interpretations when the experimental question is broad and/or poorly defined. Thus, narrowness and specificity are important aspects of a good experimental design. Yet when the experimenter follows such guidelines, he or she is often left without an easy way to attach the experimental results to an issue of social (and scientific) importance or significance. One of the most difficult jobs in science, then, is to be experimentally (and interpretationally) precise without losing sight of the "big picture." It's a difficult balancing act.

I believe that for an individual early in his or her scientific career, the emphasis should ordinarily be on the side of specificity.

Try to be scientifically conservative.

Aside from making your scientific work somewhat easier, this emphasis typically has an important positive outcome with respect to your career: By being clear and specific, you gain a reputation for being an incisive thinker and experimenter, and people are much less likely to take you to task for "over-interpreting" your data.

Integrating one's experimental results into a "big picture" scenario often requires not only the breadth of knowledge, but also the intuition, that comes with experience. Yet even the beginning investigator needs to know (and should be able to explain) how his or her study fits into the big picture, how it relates to the results of other experimenters, how it builds on and complements past and current research concepts. Putting things together, as in a puzzle, is after all what we as researchers are all about. None of us generates a complete and finished picture; most of us simply add pieces to the puzzle, and it helps to see where your piece fits.

You can facilitate this integration by approaching the interpretation of your experiment by telling a story ("once upon a time"). Think about the history of the problem, about how others have approached it, and how your approach constitutes a novel way of thinking about it. While you may never be able to end your story with "and they lived happily ever after," the story does help you realize that you're not alone in this enterprise. It helps you integrate the details of your experimental results into an interesting and (ideally) significant context that has broad outlines and intrinsic interest. And perhaps best of all, it helps you make clear the relevance and importance of your results even to the non-specialist, who may not understand the details of your experimental approach but can appreciate the coherence that a few additional details add to an already existing story.

◆ Some Words About Critical Thinking

Telling a story and practicing critical thinking are not mutually exclusive; indeed, they complement each other quite nicely. So as you make up your story line, keep the following in mind:

1. Be clear in what you say and how you say it. Particularly for those of us who habitually work with a given set of problems, it is easy to assume that others can follow along. Although a close colleague may be able to "read your mind," most people (even other scientists) will not have the same set of assumptions and biases that you do, so spell them out clearly. And remember, if your explanations are not understood, it doesn't mean your listener/reader is stupid or dense: it's your responsibility to make your explanations understandable. If you can't explain it, you probably don't understand it yourself.

2. Understand, and express, the difference between correlation and causation. Ideally, most of us are looking for causal relationships, but practically, many experiments yield only correlational information. The difference is profound, particularly when you're presenting your interpretations and conclusions.

3. Avoid circular thinking. Especially when confronted with a difficult problem, we often end up confirming ("proving") a hypothesis that is, in fact, an assumption of the experimental hypothesis or design.

4. Try to keep things simple. It's usually better to choose the simpler explanation, or the most direct way to test a hypothesis. Resist the temptation of the complex, and particularly of technical flights of fancy. While you might be able to wow your audience (e.g., colleagues, readers) temporarily, there will always be the need for substance. And the simpler the approach, the easier it will be to provide that substance.

5. Admittedly, in the real world, things are not always simple. Therefore, it is *also* important to recognize the complexities of your system. In science, the key to success is often being able to balance these conflicting tensions. One of the most important and most difficult skills to learn is when to simplify and when to work hard at elucidating the complex interactions associated with your problem.

6. Learn to differentiate between minor (e.g., technical) defects and major (e.g., structural and/or conceptual) defects. This sort of differentiation is important, not only in evaluating the work of others, but also in assessing your own experiments. It's surprisingly difficult, at least in some cases, to distinguish between a praiseworthy focus and a preoccupation with the trivial. While it remains true that "the devil is in the detail," detail in itself does not guarantee significance.

7. When you interpret your data, stay within the bounds of what your data actually say. Speculation is certainly allowable in science, but it's important that you recognize it as such and label it clearly as speculation for your listener/reader.

8. Above all, practice. As I said at the start of this chapter, thinking like a scientist is an acquired art. That capability is facilitated by repetition, constructive criticism, and a willingness to put yourself "on the line." With practice, this mode of thinking becomes second nature, something that you bring not only to your work as a scientist but also to issues in other aspects of your life.

**Real-Life Problem**: You've been working on a research problem and collected lots of data supporting your hypothesis. You therefore conclude that you've proved your hypothesis. However, when you submit your research to a journal, the reviewer objects to this conclusion. What might be the problem?

*Continued*

1. You can't "prove" something to be true on the basis of data collection. All you can say is that your studies are consistent with your hypothesis.
2. You haven't included the appropriate controls.
3. Your hypothesis involves circular thinking (i.e., you start off with a condition that requires your conclusions to be "true").
4. You've over-interpreted your data (i.e., the data don't speak directly to your hypothesis, or your experiments are too narrow to support your general hypothesis).

**Discussion**: A reviewer may be responding to any one, or all, of the above possibilities. They are all dangers awaiting even the most experienced scientist. It's helpful to get into the habit of avoiding the term "prove" and to work hard to be one's most severe critic. Over-interpretation will rarely be a problem if one practices *conservative* scientific thinking. And thoughtful experimental design virtually always involves the use of critical controls—although the question of which controls are really important for a given study may be a legitimate focus of disagreement among scientists. Detecting (and avoiding) circular thinking is sometimes best accomplished by seeking outside input; it's sometimes difficult to gain an "objective" view of the problem (and your hypothesis) if you've had a long and intimate involvement with that problem.

# 4 ◆
# How to Write a Scientific Paper

◆ What is a Scientific Paper?

Now that you've designed and carried out your experiment, you'll need to convey your findings to others in the field. The most effective way of doing so is to write and publish a scientific paper in a widely read journal. The success you have in telling your colleagues about your results, and in making the case for their significance, depends largely on how well you convey your information. Most students have experience in writing papers, which are then "reviewed" (i.e., graded) by their teachers. These school assignments provide critical background for the task of "writing up" your experiments. Many of the skills you learn at this earlier stage of your education are important for communicating scientific data in a journal article. There are, however, some peculiarities of writing a scientific paper, particularly with respect to format.

49

In general, research reports are organized into six major sections. These sections have different names (and may be presented in different orders) in different journals and across different fields, but they reflect the type of information that most scientists would like to have in learning about the studies of their colleagues. These papers, then, have the following organization:

1. *The abstract or summary.* An important part of any research paper is an initial Summary that provides a brief review of the key features to follow. Writing effective abstracts is an art unto itself—and well worth the effort, since many readers will look only at the abstracts of articles of interest, and almost all readers will consult the abstract to determine if they want to examine the article in further detail. The abstract should briefly (usually in about 200 words) tell the reader the purpose/goal/significance of your experiment, identify key methods and salient results, and give a brief statement of conclusions. If the abstract makes a sufficiently good case for the importance/interest of the article, then you might gain additional readers. If you can't make the abstract interesting, you can generally forget about readership from all but a narrow group of colleagues (usually those whose research is precisely on the topic of the paper).

2. *A statement of your ideas and hypotheses, with a brief discussion that incorporates those ideas into the history of research in the area (i.e., the background [context] for your study).* This information, often contained in the Introduction section, should help the reader understand what question you examined and why it is significant. To make that case, it is important to provide some historical background, particularly some discussion about if/how the question has been approached by others. The relevant previous work should be cited and the detailed citations listed in a reference list. This introduction is also a place where you can

comment on the "holes" that still remain (i.e., where there is a need for additional experiments and better data)—holes that your experiment will presumably fill.

3. *A description of how you did your study and how you analyzed your data.* This section, often called Methods, should present your experimental design, the methods you used in your experiments, and details about statistical analysis. The descriptions should be provided in sufficient detail so that someone who wants to replicate your study could obtain the needed information to do so from reading your Methods section. Complete methodologic detail, however, is often inappropriate when those details have been given in previous publications; in such cases, citations to those earlier works are acceptable.

4. *A clear presentation of experimental findings.* The Results section is the place to tell the reader what you've discovered in your investigations. The results should be presented clearly and completely. Figures and tables are often helpful in presenting the data and can be more effective than trying to give the results in elaborate text descriptions. The results of the statistical tests that you used to compare experimental groups (e.g., controls and manipulated sample groups) also belongs in this section. Most journals encourage authors to present their data, at least initially, without discussion about their meaning or importance. If the experiments are simple, then the Results section may be quite short. If the studies are complex (involving many experimental groups, many manipulations, and so forth), then this section may be quite extensive. The author's goal should be to provide the reader with a complete and "unbiased" account of experimental results.

5. *A critical discussion of your findings,* including interpretation of the data, particularly with respect to the questions and hypotheses presented in the introductory section. The Discussion section is your opportunity to say what

you think the experimental results actually mean and to draw conclusions from your study. Discussions should remain focused on the specific question(s) asked, and conclusions should always be limited by the actual data; that is, conclusions can be drawn only on the basis of experimental results (and certainly not on what one might like the data to mean). Limited speculation is generally acceptable in most journals, so long as it's clearly labeled as such; don't try to sell your speculation as conclusions based on nonexistent data.

6. *Additional related information.* Such information generally includes a list of citations that you've referenced in your paper (with sufficient detail to allow the reader to find a paper of interest), acknowledgements of help and financial (grant) support, disclosure of conflicts of interest, and other details required by the journal.

This outline is, as indicated above, typical for most research reports. In the current environment of electronic publication, some journals also allow authors to include "supplemental" information online—electronically linked to the main body of the paper. Such an addendum provides the authors with an opportunity to give the reader additional details (e.g., long tables, extra figures) that support and expand the key data given in the main part of the text but that cannot be included there because of the space/length limitations imposed by the journal.

There are, in addition to these standard research reports, other types of papers that follow slightly different formats. For example, you might be asked to write a review, in which you try to summarize the key information to date about a certain topic. This type of paper requires a different type of "research" (primarily into already-published articles in the scientific literature) and a different type of analysis (a critical integration of work from many different laboratories). One might consider these reviews as composed of expanded Introduction and Discussion sections as described above.

Another type of paper is a short "letter to the editor," commenting on something you read, about which you have a different (or supporting) viewpoint. Such letters are generally very brief, do not present methods, usually contain only a very brief (if any) presentation of original data, and are narrowly focused.


◆ Writing a Paper—How to Start and What to Say

Like other writers, many scientists often feel "blocked" when they sit down to write a paper. Indeed, for some (and this may be especially true of young investigators), the task of writing a paper appears so monumental that they just can't get started. As a result, experimental results pile up on their desk, never to see the light of day. This situation is regrettable on several counts. First, it may significantly hamper career progression, which often depends on publishing (and publishing often) in the scientific literature. Second, the field never learns the results of what may be important, ingenious studies. Third, because the results stay hidden, the funds that were used to support the research are, in a real sense, wasted, since the experimental data cannot be used to advance the field.

> Any way you look at it, writer's block for a
> scientist is just not an acceptable option.

The one tip I can offer is that waiting—until you have "all the data," until you have carried out every possible experimental manipulation, until every nuance in analysis has been carried out—is a sure recipe for disaster. In my experience, an experiment is almost never complete. There's always something else you can do. There's always some bit of information that would improve the experiment or make the results clearer. Given that state of perpetual incompleteness, it makes little sense to wait for completion. So don't wait. Indeed, many scientists recommend starting to write the paper before the experiment is carried out. One can usually compose the Introduction and Methods sections without the need for experimental data and then simply plug in the results when experiments are "completed." This approach not only eliminates the temptation to procrastinate but also helps the investigator focus on the goals of the experiment.

There are many ways to write a scientific report—many tricks for getting started, many mechanisms for constructing the paper, and many styles of presentation. I present a couple of options here based on my own methods for approaching this task; please note, however, that they are not the only useful approaches, nor are they mutually exclusive.

Option 1 follows the outline method. As we learned in high school, an outline identifies the major sections of the paper (above) and allows the author to insert important points in a logical order. At the start, no details are needed, just relatively broad topic headings. Such an outline then provides a framework into which you can insert details of experimental protocol, specific experimental results, and so forth. The Results section may then grow out of a series of figures, identified in your outline, that show the major experimental results.

Option 2 takes another tack: it's basically a "stream of consciousness" approach. Especially for those who have trouble getting started, it may be helpful to sit down at your computer and type out your thoughts about the problem you're working on, about the results you've obtained from your experiments, and about what these data actually mean. Using this process, you don't need to worry about order or detailed results; nothing is set in stone. Indeed, you can use this approach before you've even started your experiment—and certainly before you've begun to analyze your data. Simply make a guess about how your study will turn out. If you're wrong, you can correct yourself later, but at least you've made a start.

I strongly advise my students to start the first draft of their papers—realizing that there will be many drafts—without agonizing too much over details. The order of presentation can be changed. The methods of presenting the results can be improved. Discussion and conclusions may turn around dramatically once details are inserted. Virtually every aspect of the paper can be adjusted, once you know what message your paper will convey (i.e., the results of your experiment). For many of us (including me), that message is often not clear at the outset but develops as you become more intimate with your experiment and its results, as you analyze the data, and as you consider the many ways of evaluating your results.

Remember, too, that while a simple, straightforward report of your experiment is perfectly fine, it is unlikely to make an impact (be remembered) unless it is linked to something that the reader cares about. So tell a story—a story that someone is likely to want to hear/read, because the topic is so intriguing, because the problem is so compelling. To compose such a story, you must make a strong case for the significance of the problem and of the experimental data. And to establish such a case, you'll inevitably want to draw on the history of research in the field (i.e., provide key links to the literature), placing your work within a larger context. You'll want to show how your experiment moves the field forward, toward better understanding, greater insight, and so forth. In other words, you'll want to connect the dots, make explicit the connections that illustrate the importance of your study. Drawing attention to these relationships is part of your job, to show how your particular experiment informs (and fits into) the big picture.

Working toward this goal is a tall order. You can easily get carried away with yourself, or become so cautious that the report loses significant impact. A useful practice is to write a draft (or two), and then put the paper away for a couple of weeks. When you come back to it with fresh eyes, you'll be able to see the strengths that need to be further developed, as well as the weaknesses that need to be improved. This practice will also help you avoid one of the most common pitfalls we all experience in presenting our results—the assumption that our readers can read our minds. We all have a penchant for assuming that our readers have the same thought processes, background, and assumptions that we do.

> But in general it is best not to assume very much about your readers.

To write a paper without making assumptions, you will be forced to be clear and explicit about your ideas, your hypotheses, and your interpretations of the data. Just because something is obvious to you (e.g., a relationship between your results and your hypothesis), don't assume that that relationship is obvious to your reader. Avoid using the phrase "Thus, we can see"—unless you're really sure that the "thus" is warranted and that "we" can actually see the connections.

◆  Presentation of Data—Figures and Tables

The real nub of any research report is the presentation of data. Telling your colleagues about your experimental results is, after all, the rationale for writing a paper. While it is often possible to convey this information in words, your verbal descriptions can be significantly facilitated—or even replaced—by judicious use of figures and tables.

Many readers will, in fact, "read" a paper by going directly to the figures and tables to get the story. While there is no one way to think about using these devices, some important general considerations are worth discussing.

Tables are generally quite straightforward. They offer the author an opportunity to list, under appropriate headings, quantitative data or descriptive information that would require many laborious pages of text. Given this rationale, the data included in a table need not be explicitly recounted in the text; rather, it should be summarized with a reference to the table for the complete story. Tables are helpful ways to present complex data—but if the table itself becomes too complicated, its usefulness decreases. A reader should be able to pick out the salient bits of data from a table, not get lost in its complexity. Indeed, most readers simply skip over long complex tables.

Figures, in contrast, come in all shapes and sizes and may include photographs (e.g., microscope pictures, gels), diagrams, and graphs of different sorts. Effective figures show the key data clearly, with adequate size, clear labeling, and appropriate explanation in the legends that accompany each figure. Complex figures, consisting of multiple parts, provide the author with a key means of presenting results that would be difficult to describe within the text of a research report.

◆  Some General Tips About Writing

First and foremost, practice! Don't worry about getting your paper "right" the first time. Ask someone else—an advisor, an experienced colleague—to read your work and critique it. Use this feedback to improve your presentations. Assume that for every paper you write, you'll generate multiple drafts, each of which improves on the previous one. The intimidation factor that is so often responsible for writer's block is often wrapped up in the idea that you need to write the paper perfectly the first time. No one writes that well—although some seem much less reluctant to initiate the process than others. Even senior researchers, with dozens of publications, often struggle with the actual write-up of their experiments. Indeed, when I hand a manuscript draft back to one of my students with a lot of red ink, I assure him or her that I do the same thing to my own drafts. Rewriting, correcting, improving is an intrinsic part of the process. Papers that sound good, that are clear and interesting, are usually the result of revision and correction. It is sometimes helpful to read a paper that you think is particularly well done and try to analyze the features that make the paper interesting and accessible (aside from the actual data presented). If you are drawn to a given style or manner of presentation, dissecting the features of this style may help you apply the style to your own written work.

Effective scientific writing is, in many respects, no different from good examples of other types of literature. I list a number of general aspects of effective writing here, with the reminder that most good writers develop a distinctive style that is expressed within these general guidelines:

1. *Organize your thoughts* so that you can present them in a logical order that makes it easy for the reader to follow. Few of us can generate a coherent manuscript in our heads or pour it out in finished form as a stream of consciousness. So use whatever mechanisms that will assist you in organizing your work. Outlines are often helpful, since they can give you a snapshot of the entire manuscript, with a sense of how the parts interrelate. And they can (as can input from others) help you in reorganizing the parts so that they make more sense within your larger framework. Don't be reluctant to try out different revision possibilities.

2. *Learn to use language effectively.* Reading good literature—apart from science—can be a good way to develop a feel for the use of the language. Learn the basics of grammar, and make sure that what you write really makes sense (e.g., each sentence has a subject and a verb). Here again, there is a great tendency to assume that what you write makes sense, because you have "privileged access" to your thought process (access that your readers won't have). So it's important to ask, about each sentence, whether what you've written actually makes sense.

3. *Use easy-to-understand constructions, and say what you mean.* Short, active sentences are almost always better than long, passive, complex expressions. We as scientists often get involved in complex phrases, since very little of what we want to say is simple. There are almost always modifying thoughts and phrases that we want to add. And somewhere along the line, most of us are taught that scientists should be "objective" in their expressions, and therefore we should avoid using an active, subjective voice. We want to give the impression that our data are "interpreting themselves" (i.e., we are not introducing our personal views), and so we are likely to say something complicated like "These results can be seen to support the hypothesis that. . . ." What we really mean is "Based on these data, we conclude that. . . ." Or we might be tempted to say, "These studies were carried out . . ." whereas what we really mean is, "We carried out these studies. . . ."

4. *Work at linking ideas.* An outline will provide help in establishing a logical order of presentation, a "flow." An outline can also help the writer to generate subheadings within a given section of the paper; subheadings help make it very clear what point you're addressing in the following paragraphs and should enable you to work out some cohesion across paragraphs. We all learn in school that a paragraph should express a particular point or idea, but constructing paragraphs with that goal in mind is challenging. Even more

challenging is linking the paragraphs together so they tell a coherent story that is made up of several, related points.

5. *Read the literature and be familiar with the work in your field before you set out to make your own contributions.* You certainly don't want to reinvent the wheel, so your exploration of the literature should precede your experimental design. Then check the literature again as you write your manuscript to be sure that you've given credit where credit is due (i.e., appropriate citations). To ignore other contributions that are relevant to your story not only detracts from the significance of your own work (i.e., how it fits into the big picture) but is also insulting to those who have made those earlier contributions.

6. *Be clear on what you want your readers to remember about (take away from) your paper.* Make sure that in your story, you've pointed out the salient features of the work—improvements in technique, key new data, a new way of interpreting the results that are available on the topic of interest. You can convey this emphasis by organizational tricks: for example, "I'm going to tell you what I'm going to say, I'm going to say it, and then I'm going to tell you what I've said."

◆ How to Decide Where to Publish

Once you've designed and executed an experiment, it may be relatively easy to determine what you want to write about. It may be more difficult to decide *where* to publish your results. And since different journals have different audiences, different "standards," and different styles, the decision about where to publish may significantly influence how you write your paper. My bias, as should be clear by putting this topic toward the end of this chapter, is that you should first write your paper (at least in first drafts) and *then* decide where to publish it. You, as experimenter and author, should be in control of what you write. Having said that, I recognize that this approach may not always be realistic, and so it's worthwhile considering what factors should be weighed in deciding where to send your paper.

The most important consideration, at least in theory, is your target readership—who you want to read your paper. Different journals appeal to different groups of scientists. Is your paper of very general interest—and therefore appropriate for a journal that has a generally broad coverage? There are a relatively small number of such journals (most journals have a rather narrow focus), and these broad-audience journals are very competitive (e.g., *Science* and *Nature*). Or are your findings primarily of interest to a particular subspecialty of a specialized field? Identifying the audience you hope to reach should be a major—if not the primary—factor in deciding where to submit your manuscript. One would ideally like to publish in the journal with the best reputation that targets the most appropriate audience.

A number of more self-serving factors are also important in publication—perhaps particularly to young investigators. The most prominent of these issues are the journal's "impact factor" and its reputation for selectivity. The impact factor refers to a calculation that reflects the number of times that articles in a given journal are referenced by other authors. The impact factor is designed to provide a measure of how widely read and used the journal's articles are. It is a measureable factor—a number—that is sometimes used by committees that determine a young faculty member's promotion or suitability for grant funding. Similarly, the reputation of the journals in which one publishes is widely considered in these evaluations of promotion and funding. Although journal reputation is sometimes equated with its impact factor, it is actually a very subjective matter (and therefore quite difficult to assess when determining where to send your manuscript).

While journal reputation and impact factor are realistic concerns, neither of these actually measures how broadly a journal (or an article) is read—a matter that is now reflected in a statistic based on the number of downloads for electronically published material. Importantly, these "measures" do not necessarily determine where your paper would have the greatest effect on the field, or be most useful and appreciated. So it's worthwhile to think about a journal's readership, and the journal's breadth and focus of circulation, as well as the prestige one might gain from publishing in a "high-tier" journal.

Realistically, too, one needs to consider which journals are likely to accept your paper. Journal editors do recognize the choices their authors face. A journal that receives many papers that are considered to be of high general significance is unlikely to consider a paper that presents a narrow and very specialized finding. You as an author need to recognize the niche in which your paper fits, and identify a reasonable set of journals to which you might send your work. In addition to the journal's readership, focus, and acceptance rate, you might also want to consider, among many other factors:

1. The journal's format: Will that format allow you to present your data easily? Can you present the number of figures you think you need, and in large enough a form to be readable?
2. The journal's reputation for fair and rapid peer review
3. The quality of the journal's presentation, particularly of its figures (e.g., will photographs be reproduced accurately?)
4. The journal's use of (and charge for) color figures

All of these seemingly minor matters can influence the ease with which you can publish and the impact your paper will have on colleagues in your field.

◆ Authorship

A quick perusal of the literature in almost any field will reveal that relatively few papers have single authors. As I've indicated in other chapters, scientific research is largely a cooperative, collaborative enterprise, and the report of research studies will usually reflect the collective work of several individuals. There is ongoing discussion about how to determine whether a given individual should (or should not) be considered an author, with this discussion focusing primarily on the individual's role in the study. It seems obvious that the individual who actually writes the paper should be an author, and it seems obvious that the individual who has done the research should receive credit as an author on the paper. These roles may or may not be represented by the same individual, and in both cases more than one individual is likely to be involved.

Most journals indicate that the individuals who are listed as authors of a paper should have made a "significant intellectual contribution" to the study being reported. This criterion is helpful but rather "fuzzy." Because of this criterion, research technicians—even those who do much of the technical work on an experiment—are often not listed among the paper's authors. In contrast, individuals who may have made little practical contribution to the experiment (e.g., a department chair or laboratory head who makes critical facilities available for the research) are sometimes included in the author list (although this practice is now actively discouraged). Determining whether a given individual has contributed "sufficiently" to the study to be included as an author is generally made by the paper's senior author.

Given that most papers have multiple authors, an implicit code has developed by which a reader can make an educated guess about who did most of the work—the "first" author. (Some journals now require that the role of each author be explicitly identified.) Similarly, according to this code, the last name on the authorship list is usually the senior author, the individual in whose laboratory the study was carried out and who likely generated the funding to support the research. Negotiating one's place on this list can be an awkward business, particularly for a trainee. As a student or fellow, you would like to be recognized as first author as often as possible. First authorship is expected when you publish your dissertation research, or when you have contributed the majority of effort to the design and execution of a study. But if a trainee does not write the paper (e.g., if the lab supervisor takes over that task or assigns it to another more senior individual), or if the PI believes that someone else has made the major contribution to the research project described in a paper, the trainee may be relegated to a second or third author position.

Determining authorship position is a delicate matter. Significant confusion and disappointment can be avoided if this matter is clarified early in the writing process. It's best not to be surprised (and disappointed) to find that you are not first author when the manuscript is finally submitted.

◆ Manuscript Reviewing

As a contributor to the literature in your field, you will inevitably be asked to serve as a peer reviewer for journals that deal with your subject matter. Reviewing the work of others will be part of your job. And indeed, as you become better and better known in your field, those requests will likely increase. While it's certainly possible to decline these requests, as a member of your scientific community you will be expected to agree to review at least some of the manuscripts that are sent your way. Although the review of manuscripts requires considerable time and effort (without any significant payment), there are a number of benefits of this activity that should encourage your participation:

1. You establish a reputation as a participating colleague—and if you're good at it, you will enhance your reputation as a critical thinker.
2. It is a useful way to keep up to date on the latest advances and findings in your field.
3. You learn how to anticipate the kinds of comments a reviewer might have about your own work, and thus learn many useful lessons about how to write a paper that reviewers will recommend for publication.

Just as writing a manuscript is an acquired art, so too is manuscript review. There are good reviewers and not-so-good reviewers, and journal editors are quick to learn which are which. And just as there are "tricks" to becoming an artful manuscript writer, there are factors—that can be learned—that will enhance your skills as a reviewer. Young investigators can benefit from reading (and, indeed, sharing in the generation of) reviews written by their mentors. For example, comparing your review of a paper to the review from a more senior scientist can help you identify and distinguish major issues from minor criticisms. Among the pointers that help in review are the following:

1. Learn what the journal editors find useful in manuscript reviews, and try to make your comments helpful and relevant. Your job as reviewer is not to accept or reject a paper (that's the job of the editor) but rather to identify strengths and weakness of the manuscript. Journal editors find it helpful if you tell them whether the paper has (a) clearly stated goals and hypotheses; (b) well-described methods; (c) clearly presented data; and (d) discussion in which the conclusions are

supported by the results. Recall that these are all issues that you, as an author, should be aware of as you write your manuscript. Don't waste the editor's (or authors') time with trivia. In particular, if asked to make a recommendation (to the editor) about the paper's suitability for publication, be sure your recommendation is consistent with the comments/critique you've provided for the authors.

2. As you identify weaknesses in a paper (e.g., experimental design flaws, inappropriate methods of analysis, confused data presentation), try to be constructive in your criticism. It's not your job to tell the authors how to do their experiments. It *is* your job to ask questions and identify issues that illuminate the manuscript's shortcomings.

3. Be clear about your recommendations for revision. Your comments will often serve as the basis for the authors' revision of their manuscript, so the clearer you are in your critique, the more helpful your comments will be for the authors.

4. This process of evaluating manuscripts is called "peer review." Remember that you're dealing with colleagues. Put yourself in the author's shoes. It's fine to be critical to help the author generate a better manuscript, but don't be critical just to show how smart you are.

◆ The Changing Landscape of Scientific Publication

No aspect of the scientific research field is changing more rapidly than science publishing. While many of us continue to cling to the old and established methods of publishing in peer-reviewed journals that send out monthly (or weekly) print issues, advances in the way we can share information electronically have created a new set of alternatives. These alternatives now present authors (as well as journals) with several important choices when considering how to share experimental results. For example, it's now possible to disseminate your data easily, inexpensively, and widely simply by posting experimental results on your personal website. You can write the usual research report, but instead of submitting it to a journal for review, you can let your colleagues know that the data can be viewed online (and provide the relevant web address). This method of personal publication avoids the hassles of manuscript formatting (for a particular journal), peer review, and manuscript revision, as well as any expenses associated with journal publication (e.g., page charges, charges for color figures).

While "easy," this approach has a significant drawback: your readers will know that the paper has not been "reviewed," and therefore may be skeptical (or at least uncomfortable) about the quality of the data. There are many arguments about the usefulness of peer review (e.g., some critics point to the fact that reviewers often differ radically in their assessment of a manuscript), and this book is not the appropriate place to discuss those issues in detail. Suffice it to say that most scientists believe that peer review is a critical step in publication and are likely to give little attention to papers that are not reviewed. At least for the immediate future, then, some form of peer review appears a significant step in publication.

A more important movement in modern publication is "open access." The idea here is that research should be widely and freely disseminated. This open-access approach not only is consistent with the spirit of scientific research but is also mandated by the fact that much (most) research is supported by public dollars (or pounds or euros). Given such widespread public support, many have argued that the results of these studies should be freely available to anyone who has the interest in reading about them. In the current publication model, access to published data is somewhat limited, at least initially, to those who subscribe to the journal in which the paper is published—even if the material appears both online and in print (which is now the norm for most scientific journals). A number of imaginative and growing alternatives have sprung up to support the open-access movement, including new journals that are freely (electronically) accessible to anyone with Internet capabilities. These journals provide a peer-review assessment and are like the current traditional journals, with one major difference: the cost of publication is borne not by the journal subscribers (there are none) but by the authors. For you to publish in such a journal, you must pay a fee.

There are other important issues affecting scientific publishing. Given the speed with which new results are being generated, there is increasing pressure for more rapid publication. The peer-review process is often quite time-consuming, and authors are rightly seeking venues for their work that will allow them to publish quickly. In addition, there is increasing scrutiny of "ethical conduct" issues associated with publication, including accuracy of data reporting and avoidance of plagiarism (i.e., using data, ideas, or even language of others without citing the source of this information). Concerns also extend to accurate and complete identification of funding sources, giving appropriate credit to those who have participated in the research and manuscript generation (and not crediting those that didn't contribute), and disclosure of conflicts of interest.

There are obviously many issues to think about when you write a scientific paper. The process and the concerns go significantly beyond simply putting words on paper (or into a computer file). Much of the process is not intuitive. Take the time to learn about, and master, the various aspects of manuscript generation and publication: the effort will pay off in the long run.

**Real-Life Problem**: You've been working on a research project and are now ready to write it up for publication. You can't, however, get started on the writing process since it's unclear to you if there's enough material (sufficient data, significant results) in the potential paper to justify publication. How do you make the decision?

1. Write a rough draft and ask a mentor or colleague to read and comment on it.
2. Generate a set of figures from your data, and then see if you can write a paper around these figures.
3. Write a draft, put it aside, and then come back to it to see if you've told an interesting story.
4. Write something up and submit it to a journal, with the philosophy that the worst that can happen is that it's rejected.

**Discussion**: The first three of these choices are reasonable and useful. Writing something and then asking for input from your mentor and/or colleagues is always a good way to start. And as I indicated above, coming back to a draft with fresh eyes is often revealing. Many authors prefer to generate their data figures first and then write a text that describes the results presented in the figures. This approach provides a solid focus and is particularly good for clarifying "upfront" the data that will be described in a manuscript. Personally, I would discourage submitting something to a journal that you're not sure is ready for publication. By doing so, you present a

*Continued*

potentially suboptimal picture of you and your work to other scientists in the field. So it's really not the case that the worst that can happen is that the paper's rejected: the worst that can happen is that your colleagues form a negative opinion of your work—that is, your reputation suffers.

**Real-Life Problem**: You're completing your dissertation research, which is an aspect of a larger research program in your mentor's laboratory. You've written up a good deal of your study, but your mentor now determines that the work should be incorporated into a major paper from the laboratory, with a more senior scientist as first author. How do you handle such a "surprise"?

1. Discuss with your mentor the importance of having a first author paper based on your dissertation work.
2. Refuse to share your write-up and data analysis with the newly assigned first author.
3. Accept the PI's decision and be glad to have your name on the paper.
4. Write to the editor of the journal to which the paper is submitted and complain that you have not approved of including your data in the submitted manuscript.

**Discussion**: This tricky problem is unfortunately not so unusual. It would, of course, have been preferable to discuss with your mentor, at a much earlier stage, the publication of your dissertation work. But given that we are often not so forward-looking, it's important to develop some strategies for dealing with this type of occurrence. The first option—discussing this decision with your mentor and lab PI—is the obvious first step. Such discussion may well have the desired effect, since the PI may just need a "reminder" about how important it is for you to be a first author on your own dissertation work. But given a negative outcome, the Ph.D. student has little choice but to accept the decision and learn from it. Inclusion as an author— even if not the first author—on a significant/major publication is not the worst of possible outcomes. Refusing to share your data is not an option (the data don't really belong to you—see the section in Chapter 12 on intellectual property). Finally, especially for a junior scientist, complaining to a journal editor will most likely simply contribute to a reputation as a "trouble-maker."

# 5 ◆
# Giving Presentations and Talks

◆ The Verbal Presentation—A Fact of Scientific Life

In the previous chapter, we focused on the process of writing and publishing research reports as a key aspect of a researcher's professional life. So, too, are presentations in other formats. As is the case for the written research report, other types of communication are all about conveying your experimental data, and your ideas about your research, to your colleagues and other interested parties. A verbal presentation before large or small groups—be the presentation only a relatively informal discussion in front of a poster, a seminar presentation within your own laboratory, an invited talk at another institution (e.g., a "job-talk"), or a formal lecture in front of a large audience—is an essential tool in your scientist toolbox. Whether or not you're comfortable in this role of presenter, you need to master the craft.

Several aspects of these presentations are worth considering. At the core, of course, is the issue of effective communication. Presenting research results is an important mechanism not only for disseminating your data and ideas but also for eliciting direct feedback that can help you refine and integrate your research efforts. These talks also provide investigators with an opportunity to make their communications more "personal." Although your personal style certainly shows in a written report, the verbal presentation is an even more effective forum for presenting yourself as an individual to your colleagues. Through such talks, other scientists get to know you; these presentations are therefore important for your career in research.

> Presentations before groups of people are
> performances.

For those who have the gift of showmanship, or who like to engage in discussion and debate, these presentations offer a forum in which to engage these "nonscientific" urges. Some people love it—the opportunity for storytelling, a stage for the dramatic, an audience for the comedian. But talking in front of a group of people can be a real challenge for others. For those who are "verbally challenged," who are shy and would rather spend their careers in isolation in their laboratories, this requirement for conveying their experimental results to an audience is anything but inviting. Yet, in the present scientific society, these presentations are unavoidable. So love it or hate it, you need to get good at it.

As with writing a paper, the key to success in public presentations and talks is practice. Some people are "born" entertainers and take to this public life easily; the rest of us need to work at learning presentation skills. You can learn to be a good speaker, even if you're not a showman.

Here are a few pointers that should be helpful, whatever your reluctance to give talks might be:

1. *Know what you want to say.* Practice your talk in front of a friendly audience (e.g., at a lab meeting). Perfect your presentation so that you know exactly what you want to say and how you're going to say it.

2. *Know your material.* The fear that leads to extreme nervousness and reluctance to present often comes from the idea that you will be "found out"—that people will "discover" that you don't know what you're talking about. If you know your material, this fear should not be an issue. Realize also that if you're giving a talk on your particular area of research, odds are good that you know much more about the topic than anyone in your audience.

3. *If you're afraid you'll forget what you want to say, write out your talk.* Use written notes, or use your slides to guide you smoothly through your talk (more of this below).

4. *Generate energy and enthusiasm for your subject matter.* If you sound like you're excited about what you're saying, your audience is likely to share that excitement. If you sound like the topic bores even you, then there's little hope that your audience will find your presentation very interesting.

5. *Engage your audience.* Your interest, and their interest, can be maintained if you can establish some sort of dialogue. This interaction can be verbal or can be a function of feeding off other types of responses from your audience (e.g., facial expressions, laughter, applause).


◆ Keys to a Good Talk

There are as many ways to give a talk as there are individuals. Each presentation carries with it the character of the speaker, as well as the impact of the data and concepts being presented. However, you should consider several basic aspects of presentation in generating the best possible talk for you:

1. *Know your material.* I know I've already made this point, but it's worth making again. The anxiety about verbal presentations can be significantly alleviated if you feel confident of your command of the material. Moreover, that command is always obvious to the audience, and they will therefore "relax" in the expectation that you'll tell them what they need to know. That confidence and command make for compelling presentations.

2. *Organize your material in a logical, easy-to-follow sequence.* You don't want to make your audience work too hard to figure out what you're saying. Further, if one point follows logically after another, the points you're trying to convey are clearer and easier to remember—not only for your audience but also for you.

3. *Be clear about the salient points in your presentation.* If necessary, state your goals at the start, and then repeat these key points at your conclusion. In other words (as I indicated in a previous chapter), say what you're going to tell them, tell them, and then say what you've told them.

4. *Tell a story.* By this phrase, I mean not only that you should present your major points in a logical order, but also that the presentation should "flow" in a way that gets the listener from here (the starting point) to there (with the conclusions that you want your audience to take from your presentation). A good presentation should be an enjoyable journey, with a starting point and an ending. Take your audience through whatever thought processes or experimental sequences are necessary to explain why and how you did your study. Doing so will help your data "make sense." It will also make your experiment understandable not only to your colleagues who are expert in your field, but also to other interested listeners.

5. *Don't present too much material.* We scientists tend to try to say more than we should—perhaps because the material is so complex—and in the process often lose or confuse our listeners. If you make your presentations reasonably simple, your talks will inevitably be clearer. And limiting yourself to a few major points (or experimental results) will be a great help in restricting your talk to its allotted time. Sticking to your scheduled time will be appreciated by both listeners and organizers.

6. *Engage your audience.* Yes, I know I said this before, but it's a good trick. Remember, the whole point of these presentations—seminars, lectures, whatever—is to get your audience (e.g., fellow researchers, students, and fellows) to think about your ideas and/or data. If you get them involved during the talk, you'll have them hooked. So try being provocative—even more so than you would normally be in a manuscript. Challenge generally held assumptions. Try to convince your audience that your study is important, adds significantly to the field, is not simply a "me-too" investigation. Encourage questions, and take them seriously (even if you think they're silly or that the answer is obvious). Be prepared to engage in discussion, and organize your talk so as to leave time for such interaction. As you engage in give-and-take, be respectful and acknowledge useful contributions. Above all, don't be a know-it-all—and don't be afraid to say "I don't know."

7. *Work hard on your slides* (or your poster). The way you present your data visually does make a difference in the way people respond to you and in how convincingly you convey information. The key points for effective research slide presentations are perhaps obvious but bear repeating. First, keep the slides simple. Be clear about what point you want to make with each slide, and keep the slide focused on that point. My bias is to avoid making the *design* the focal point of the slide; it is far too easy to distract your audience from what you want to say if you attempt to be

too imaginative with your slide set-up. Avoid too much text, and make sure that the text you do use is easily visible, even to those sitting at the back of the room. Similarly, your figures, diagrams, and tables should be simple and easy to understand. Second, work on slide *order* so that you can move forward (or backward) easily and logically. If you do this well, your slides will help you remember what you want to say. Slides should help you tell your story and therefore should be arranged with an introduction, a central section focusing on your experiment and its results, and a conclusion section. If your slides are logically arranged, questions that require jumping forward and backward are easy to address. Finally, restrict the number of slides you plan to present, so that your talk is focused, accessible, and respectful of your audience's schedule. In preparing their talks, many practiced speakers will estimate the slide number based on the scheduled length of their talk—for example, one slide per minute. But you'll need to determine that algorithm for your personal style—including in the calculation such variables as how much information you tend to include on a slide, how fast you talk, how complicated the concepts are, and so forth. Work toward finishing your presentation on schedule, leaving ample time for questions and discussion. As I've noted above, the organizers of your seminar or lecture, and your audience, will thank you for that consideration.

◆ Elements of Style

Every scientist has a unique style (including a speaking style) that reflects his or her personality and approach to the research endeavor, so spend some time thinking about your style. Certainly, for a young scientist, this style will evolve over time and will be influenced by mentors and colleagues as well as the individual's personality. This style issue is often taken for granted, but your effectiveness as a speaker is certainly something that can be developed with practice and effort. My experience is that you're generally "stuck" with some basic aspects of your personal presentation style, and while it is possible to make improvements and modifications, it is useful to recognize this basic starting point and develop it in the most effective way possible.

Effective presentations involve keeping the attention of the audience on the points you want to make. You can do this by entertaining them, engaging them, and/or making the material relevant to their interests. Not everyone is an entertainer or showman, and not everyone can be funny. Some of our colleagues enjoy the showmanship aspect of research and are comfortable telling jokes or stories that hold the audience. Some are much more comfortable with a formal, straightforward lecture presentation. And some do best when talking (or imagining they are talking) to small audiences in an informal manner. It helps to be aware of your personal propensity, since your comfort level will often determine the success of your presentation. It will also help you to determine what types of audiences you're interested in addressing, and in what types of formats. For example, if you enjoy the give-and-take of discussion, then smaller seminars are obviously a better choice than large lecture halls. One doesn't always have this choice of venue—but you do have some choice in your manner of presentation.

Your reputation as a speaker will be a part of your reputation as a scientist. The frequency with which you're invited to present your research at other institutions or at large professional meetings will be determined, at least in part, by how well you convey your story. No matter how good your research is, if you don't present it well, people outside your immediate laboratory will be reluctant to ask you to visit and/or present. An appealing speaking style will therefore facilitate your ability to interact with others in your field, help you meet new colleagues, and establish useful collaborations. Your style doesn't need to be polished, but it does need to convey your data clearly, to transmit your message, and to reflect your enthusiasm for your research.

◆ Selling Yourself

In the real world, there are two major aspects of scientific presentations. The first, discussed above, is the effective presentation of experimental results. The second, which receives much less attention in traditional training environments, is "selling" yourself. What does that mean? As I indicated previously, establishing a reputation as a careful, insightful, imaginative investigator is an important goal of every scientist. Thus, you need to "sell" yourself to your colleagues with that goal of reputation in mind. While you can do this sales job to some extent through your written work, most people form their impressions on the basis of personal interactions. How you present yourself in front of an audience, therefore, will contribute significantly to the impact that you—and your science—make on your colleagues. The quality of data presentation will be only a part of this impression.

These same issues are in play not only in lecture presentations but also in other types of "personal" interactions, whether they involve informal discussions (e.g., in lab meetings), visits to other institutions and laboratories, or formal job interviews. This book is not intended to provide guidelines or instructions about "how to interview for a job." What I would like to emphasize, however, is that the same features of effective talk presentations—interesting ideas and experimental data; command of your results as well as familiarity with relevant studies in the field;

confidence; clarity of presentation/expression—are key features of virtually all your professional scientific interactions. The style with which you present yourself is yours to determine. It is unique to each individual. It helps to be aware of that style so that you can refine it and become a more effective communicator.

◆ In the Audience

Just as the manuscript preparation enterprise has a "mirror" process that encourages/invites questions and criticism (i.e., manuscript review), so too does the stand-up verbal presentation. In the latter case, this process of critique is usually quite informal. For example, most seminars (and many formal lectures) expressly allow time for the audience to ask questions and make comments about the research that has been presented. As a member of this collegial research society, you should feel free to join in. Often you'll want clarification of a specific issue. Don't be shy about asking questions—if something is unclear to you, it's likely to be unclear to others as well. Asking for further explanation, or even providing defense of a particular point of view, is part of the scientific process. You might also want to offer relevant observations from your own work. This type of information is often useful in elaborating and complementing the presented work and can be quite valuable both to the speaker and to others in the audience. In this process, avoid the temptation of trying to show how smart you are. Phrases like "I've already done that" are sure to elicit the ire of your colleagues. But do contribute in a constructive way. Remember, too, that any type of public discussion provides you with an opportunity to introduce yourself—scientifically—to others in the field.

**Real-Life Problem**: You've been invited to give a talk on your research at a large society meeting. Although you've given several talks in small seminar environments, presenting before hundreds of people scares you. How should you prepare?

1. Decline the invitation—and others like it.
2. Convince yourself that this presentation is no different from the talks you've given in the past, and prepare similarly.
3. Write out your talk and practice in front of friends and colleagues.
4. Seek professional help—with speech preparation, with slide preparation, and even with the psychology of presenting before large groups.

**Discussion**: In order of practical usefulness, the third option is probably the best, at least for starters. Practice in front of a friendly audience is invaluable. And especially as you're getting started, writing out what you want to say can be very helpful. But all the other alternatives may, under some circumstances, be appropriate. For example, approaching a "big" talk as though it were simply a presentation in front of your lab group could be quite effective—if you can actually convince yourself of that similarity. If speaking in front of an audience is particularly frightening, seeking professional help may be a good answer. That help can come in the form of counselors who focus specifically on public speaking, or even in joining an organization like Toastmasters. If public speaking remains an unpleasant experience, even as you become more senior and practiced, the option of turning down invitations becomes a real one—but only after you've pretty much established your reputation in the field.

# 6 ◆

# How to Compose/Submit Grant Applications

◆ Grants Are the Lifeblood of Scientific Research

The world of the research scientist revolves around grants that provide the funding to carry out experiments. Virtually every research project, whatever the field and whatever the level, is supported by some type of grant mechanism. Research costs money, often a lot of money, and certainly more money than scientists can provide themselves. Research grants are the means through which scientists can pay salaries (sometimes their own, as well as salaries for research assistants, students, and fellows), buy equipment (ranging from a few hundred dollars to many millions), pay for animal subjects and reimburse human volunteers, and provide essential supplies for the laboratory. Grants may pay for travel, consultants, contractual arrangements with other laboratories or institutions, rents for laboratory space, and virtually anything else you might imagine. They are, without exaggeration, the lifeblood of the research community.

In most fields, and particularly in today's scientific environment, grants are highly competitive. There are limited funds available, and there is a rapidly expanding pool of applicants, all of whom think they are needy and worthy. Depending on the field, the opportunities to obtain funding, as well as the competition for the funds, may vary quite dramatically. But regardless of your field of work, there is no skill more valuable than that of grant-writing. A scientist can "do science" without a faculty appointment or a high-paying corporate position (e.g., as a research technician), but the actual practice of research always requires funding.

In addition to learning how to write grant proposals, it is also important to discover, given your field and your research interests, the sources of available funding. There are funding sources for large and small needs, for fellowships and for equipment, for applied and for theoretical work. The money may come from institutional sources (e.g., your university or company), from private foundations and philanthropic organizations (e.g., the Ford Foundation, the Guggenheim Foundation), from private industry (e.g., pharmaceutical companies), or from the government (e.g., the National Science Foundation, the National Institutes of Health, the Department of Defense, the Department of Energy, as well as state and local grant programs).

Different granting agencies have different objectives associated with their funding programs. Some are very broad in interest, whereas others are narrowly focused on a particular topic, disease, or field. Some are designed to provide stipend support for the applicant, while others are directed exclusively toward equipment needs. Some are meant to foster interdisciplinary and/or interlaboratory collaboration, while others are meant to provide support for the individual investigator. There are support mechanisms that are clearly oriented to the large, active laboratory that has already established a track record. And, fortunately, there are also start-up grants and programs designed exclusively for the young investigator. Young scientists can, with proper homework (facilitated by Internet searches), discover the sources of support that match their experience, financial needs, track record, and area of interest.

Different jobs, and different funding mechanisms, will determine how often you need to go through this process—e.g., every year, every 4 years. Various factors will also determine how competitive the application process might be (depending on the amount of money available for that particular program and/or the number of other investigators in your field who are interested in a piece of that funding pie). The more competitive the program, and the more money at stake, the more complex and demanding a particular application format is likely to be. And of course, there are other determinants of the specifics of the grant application process. For example, government agencies tend to require that you fill out lots of forms, whereas some private funding processes ask only for a couple of pages of narrative and a simple budget.

The determination of successful funding is almost always based on the scientific merit of the proposed research. Yes, your name—your reputation, your track record, your prominence in the field—often counts for something. But that name recognition usually provides only a foot in the door; the door is opened by the creativity and excitement generated by your research proposal.

One variable that is always important for the success (or not) of your application is the composition of the group reviewing your proposal. Members of such review groups are generally experienced researchers, chosen (by the granting agency) for their expertise in the general area of the grants to be reviewed. Depending on the charge to the review group, members may represent quite disparate disciplines, so that the panel can consider applications that span different areas of interest. In most cases, there is a "peer review" philosophy underlying this process; that is, your application will be evaluated by individuals who are your professional peers (you, too, will eventually be asked to serve in this capacity). In most cases, it will help your cause if there is someone on the panel who is involved in, or at least interested in, your particular specialized area of research. In some cases, there's nothing you can do about the composition of the review panel. But sometimes, especially when dealing with large granting agencies (e.g., NSF or NIH), you can actually search out a review group that includes individuals who are familiar with your area of research and ask that your application be funneled to that review group for consideration. As a rule—assuming you are submitting a scientifically strong application—it is better to have your application reviewed by someone who is an expert in your field and can appreciate the strengths of your proposal. You might imagine that you could "snow" reviewers who are not familiar with your area of work, but if the application is read by people outside your field, it is much more likely that they will not understand (scientifically or thematically) your application and will score it less enthusiastically. So it pays to search for the most appropriate funding agency and/or grant mechanism for your particular idea/proposal. It is, after all, your responsibility to match the topic of your research (and the size of your proposed budget) to the appropriate review agency.

◆ The Basic Format

All granting sources tend to develop their own application formats and requirements. Despite this range of requirements, however, most application forms ask for the same type of information. The basic requirements can be summarized as follows:

1. Information about you, the applicant (particularly important in applications for fellowship support). The "biographical sketch" is designed to provide the information that will help the grant reviewers determine your qualifications; it generally includes details about educational background, positions held, and technical expertise. This part of the application is also likely to ask for some information that will reflect your track record, including honors and awards, invitations to present your work at national and international conferences, current and past grant support, and publications (e.g., research reports in peer-reviewed journals, edited and monograph books). The publication record is perhaps the most important of these categories; in addition to simply counting the number of publications, reviewers may consider the number of papers on which you've been the primary (first) author and the quality of the journals in which you've published.

2. Information that will provide a picture of the environment in which you'll be working. This category will include details about the facilities available for your research, including lab and office space, equipment (both within the investigator's lab and available as shared facility), and support facilities (e.g., expensive major equipment, specialized resource centers). This section may also involve a description of your collegial interactions (e.g., information about your collaboration and collaborators, technical support personnel, students, and fellows).

3. Information about your proposed research. This category is ordinarily the most important. Your description must be clear and must emphasize the exciting and innovative

nature of your proposed research. Often, the experimental proposal is divided into the following categories:

a. A statement of your specific aims/goals. This section provides you with an opportunity to state, clearly and succinctly, what you want to do (your goals), why you want to do it (rationale), and the approach you'll take to achieve these goals. In some ways, this section (along with a general abstract that often reflects these specific aims) is the most important part of the proposal. This brief outline of your experimental plan gives the reader his or her first impression of you and your intended work. Indeed, this section may be the *only* part of the proposal that some of the reviewers will read. It pays to make a good impression! Your description can take various forms, but I've found that it often helps—not only the reviewer, but also you in thinking about your intended research—to express your goals in terms of hypotheses to be tested. You can then briefly explain what methods you'll use to test the hypotheses, give a brief statement about your predicted results, and provide a sense of the conclusions that you think these experiments will allow you to draw.

b. A discussion of the experimental problem that provides a context for your studies. Such a background section should show the reviewers that you are familiar with the field, that you understand the relevance of previously published studies for the problem you are addressing, and that your experiments will advance the field beyond the status quo. Try to present a balanced account that hits the highlights. Within the limited page restrictions of a grant proposal, there's no way that you can write an exhaustive review of the literature. Rather, identify that work that is particularly relevant to the hypotheses you'll be testing. And make sure that your effort to provide a context emphasizes the importance and significance of the problem and shows how your work will provide an important piece of the puzzle. Toward this end, it is useful to address what you know to be the particular interests and goals of the granting agency. In some cases, your application will be submitted in response to the granting agency's "program announcement" or "request for applications" on a particular topic. Don't be subtle about the relevance of your work to the interests of the funding agency. Don't assume that the scientific reviewers and agency representatives (who may not be scientists) can read between the lines.

c. A description of research from your own lab (and from the labs of your collaborators, if relevant) that has led up to this application and provides a basis for your experimental proposal. This section should not only present relevant "preliminary data" but also show that you and your colleagues have the technical expertise to carry out the experiments in your proposal. In short, this section should show the feasibility of the proposed research and, if possible, provide intriguing, seminal preliminary results that "hook" the reader with their potential significance.

d. Details about experimental procedures—the methods to be used and the experimental design that you'll follow. This section is the "meat" of the proposal and will usually include several components. First is a description of the experimental design itself. It's often useful here to restate your hypotheses, and then give a clear account of how you will test each of those hypotheses. You should include sufficient detail so that the reviewer can determine exactly what you will do (e.g., give the number of subjects/samples you expect to test, provide specific treatment parameters). Remember, a description of experimental procedures/

design is not the same thing as a description of experimental methods. The latter should also be given in detail, either separately or integrated into the description of procedures. You may also want to include an explanation of why you've elected to use the chosen procedures. You will definitely want to explain what results you expect to obtain ("predictions"), how you will analyze them (e.g., statistical methods, determination of significance), and how these results will address your hypotheses. In most grant applications, it is also useful to discuss the potential problems you might encounter and how you will deal with them. Remember, there is no such thing as a perfect experiment, and reviewers generally recognize this fact. It's invariably better to let the reviewers know that you recognize the potential problems with your study, and have thought about how you might deal with them, than simply wait for the reviewers to ferret out these problems. Some weaknesses in a study may be forgiven if the reviewers believe that you've thought carefully about your experiments and have alternative approaches in mind if your initial attempts are not successful.

4. Information about the costs associated with the proposed research—your budget. You will be asked to provide details that show not only an estimate of total costs, but also how the costs will be divided up. How much do you expect to spend on salaries? For whom? How much of each individual's time will be spent on this research project? How much will you spend on "supplies" (e.g., reagents, animals)? What funds are needed to buy equipment? Are there costs associated with contractual arrangements? What about funding for your collaborators? All of these costs should be identified and justified (why this service or supply is needed) within the context of the particular study under consideration.

For many grants, there is often another category in the budget. This item consists of a charge that your institution—university, company, hospital—has negotiated with the granting agency to cover the costs of maintaining the facilities in which you work. While this cost is a real part of the application, it is rarely a number over which you have any control. In a typical application sent from a university to the NIH, for example, the "indirect" costs may be on the order of 50% of the cost for "direct" expenses (e.g., salaries, supplies, equipment). So if you ask the granting agency for $100,000 to carry out your study, the actual request might be for $150,000, with $50,000 going to your institution.

◆ Tips and Cautions

That's the nitty-gritty of most grant applications. It sounds pretty straightforward, and it is, but writing successful grant applications takes attention and practice. In this context, there are some procedures that will help optimize the chances that your application will be read with approval by the reviewers. While certainly not any guarantee that your proposal will be funded, here are a few tips—and cautions—that might help.

1. Keep it simple! The most common problem in grant-writing—for both the novice and those with considerable experience—is proposing too much. The reviewer comment "this application is overly ambitious" is very common; in contrast, it's rare for a proposal to be criticized for being too modest in its goals. Remember, the people who review these grants are investigators who do the same kind of work that you do (or are proposing to do). They're pretty savvy, and they'll have a pretty good feel for what can be done given the technology, the funds requested, and the time available. So be realistic in your goals.

2. Almost a corollary of the point above, make an effort to write clearly. Stay focused on your goals. If the reviewers can't figure out what you're proposing to do (if it's too complicated, or if you haven't explained it clearly), you're sunk.

> It's your responsibility to present your proposal in an understandable and convincing manner.

If the reviewers miss something or misinterpret a part of your proposal, it's usually your fault. Once again, reviewers are actually pretty smart and perceptive. While they do sometimes make stupid mistakes, a good working hypothesis is that they won't—which lays the responsibility for making the case squarely in your lap.

3. Although it can be argued that the key aspect of a proposal is the "soundness" of the experimental design, in many cases (and certainly for some grantors) it's the significance of the work that makes or breaks the proposal. It's up to you to make a clear and convincing case for the significance of what you plan to do. Even a well-constructed set of experiments, using elegant state-of-the-art techniques, is not particularly appealing if it appears to be simply "busy-work." It pays to think about the mission of the sponsoring agency when you write the section on significance, and tailor your experiments—their goals, predictions, interpretations—to fit the agency's interests.

4. Try to avoid the "Catch-22" of preliminary results. The "catch" goes something like this. To write a fundable grant, you need to present preliminary data that support your hypotheses and show that you can do the proposed studies. However, to generate those preliminary data, you need funding. How can one avoid this circularity? Some investigators make a habit of applying for funding that will cover primarily studies that they've already carried out—that is, they stay one step ahead, doing

experiments that they expect will be key to their next grant application with the funding from their current grant. That strategy works well if you are well funded and have a little "extra" with which to pursue these future studies. Another approach is to seek funding mechanisms that don't require extensive preliminary data. Some granting agencies explicitly recognize the "catch," and many have developed pilot grants (usually for small sums) for collecting preliminary data that can then be used in support of a major proposal. There are also grant mechanisms that directly target high-risk proposals. With such grants, the funding agency recognizes that there is little preliminary evidence for anticipating a given result (or even the technical completion of the proposed experiments), but also sees that if the results of the proposed experiments are actually realized, they would be extremely important. In other words, although the odds are low that the proposed experiment will actually "work," its potential significance makes it worth the risk.

5. Finally, in constructing your grant budget, the general rule is to be realistic. That doesn't mean asking for the bare minimum. Reviewers will know pretty accurately how much it will cost to carry out the experiments you propose, and they'll be suspicious if you ask for an unrealistically low amount. Ask for what's needed. Try not to inflate your numbers: "padding" the budget always irritates reviewers. But don't underestimate the resources you'll need (e.g., numbers of subjects, salary for an expert researcher associate, costs of analyses) to do the experiments completely, accurately, and expertly.

♦ Grant Revisions and Resubmissions

Many—probably most—grants are not funded the first time they are submitted. For example, at the NIH, where I have gone for most of my grant support, the current funding rate of the Neurology Institute is about 10%; that is, for any given review session, only one grant in ten will receive funding. That number varies with the agency (even at NIH, some institutes are funding nearer the 20% level) and the funding environment (at the Neurology Institute it's been as high as 25% in the past). But clearly, even if you write an excellent proposal to NIH, the odds are good that it won't be scored high enough to obtain support the first time it's reviewed. While the odds may be better at other funding agencies (e.g., private organizations), one can never assume that a grant application will generate the requested money on the first try. Given that reality, most granting agencies, and certainly the larger ones, provide applicants with an opportunity to resubmit their proposals after they make improvements based on the criticisms and suggestions of the reviewers, as well as by incorporating into the grant new insights (more preliminary data) relevant to the application's goals. Especially in today's research environment, one should expect to resubmit grant applications at least once. There is no shame in resubmitting a grant application. And as I indicated at the outset, one of the "required" characteristics of a research scientist is perseverance. Research scientists should expect some rejection in this aspect of their careers. However, assuming the proposed research has merit and significance, sooner or later the investigator is likely to be successful in obtaining funding. Even the experienced investigator, who has applied for (and received) grant funding for many years, must periodically go through the exercise of revising and resubmitting grant applications as an expected/ongoing part of doing research.

What's involved in revising a grant application? First and foremost, read the reviewers' critiques carefully, take their comments seriously, and attempt to address these criticisms directly. You may not agree with the criticisms, but it is counterproductive to assume that the reviewers "just didn't understand." Angry or flippant responses to reviewer comments—even if "justified" (i.e., the reviewer did miss something important, or misinterpreted something obvious)—rarely change a reviewer's opinion, and certainly don't make any friends.

> Remember, the reviewers are your peers—and this
> is a peer-review process.

It's likely that one or more of the reviewers who saw your initial submission will be involved in the review of your revised application (that's simply the way review groups work, a mechanism that attempts to avoid putting the applicant in "double jeopardy"), so there's no question that you're more likely to have subsequent success if you respond carefully and respectfully. That means:

1. Respond in detail to each point made by each reviewer.
2. Make changes that you agree are appropriate. That may mean adding new preliminary data, explaining in more depth how you will analyze and interpret the results, omitting specific aims that are heavily criticized and are not necessary for the general goals of your research, and adding new aims that will address issues of reviewer concern.
3. You may vehemently disagree with some criticisms and suggestions. You are not bound to agree with all the reviewers' critiques. But for those points of disagreement, provide a carefully reasoned argument, citing relevant data or reinforcing your statements of significance. This "rebuttal" can be done respectfully, recognizing that there may be other viewpoints but emphasizing the need for your particular approach.

Keep in mind that you will rarely win an argument with a review group if you base your response on simple difference of opinion (e.g., whether a particular topic is interesting or important). Remember, also, that if the reviewers have missed or misinterpreted your point, it's probably because you haven't presented it clearly, not because the reviewers are stupid. If you respond fully and carefully to the reviewers' comments, then your revised application is likely to be more successful. Depending on the research environment (i.e., the availability of grant dollars, which tends to oscillate slowly over the course of many years), it may take several tries to get a proposal funded.

If, however, your revisions don't result in increasingly better reviews (and scores), you will need to do some hard thinking about whether to continue in your pursuit of that particular research topic. There is an aspect of the research business that involves learning when to "cut your losses" and going on to something that is likely to be more productive and gratifying.

Some applications, to some granting agencies, will simply not be funded. For example, in some cases, the review group simply isn't interested in your topic or doesn't think your research is relevant to the mission of the granting agency. In such cases, it's hopeless to plead your case. Unfortunately, reviewers don't always tell you that they're not interested. For some reason, they think it's "kinder" to find numerous faults that render the application unfundable. You need to learn to "read between the lines" for indications of the reviewers' interest and enthusiasm for your topic. Reviews that indicate concerns about the novelty or significance of the hypothesis, or about the feasibility of the methods, may actually be telling you that the reviewers simply don't like your ideas. And certainly, if similar comments are repeated in response to grant resubmissions in which you think you've addressed the reviewers' critiques, you would be well advised to "see the writing on the wall." If there appears to be little interest in your question or topic, it would be much more productive to look for a more sympathetic funding source. Alternatively, in some cases it may be possible to reorient your research goals to address the funding agency's concerns while still carrying out the research in which you're interested.

◆ Effective Grant-Writing is an Acquired Skill

As I've indicated in previous chapters about manuscript preparation and talk delivery, grant-writing is a skill that can be learned, improved, and refined with practice and experience. That doesn't mean, of course, that as you get older (and wiser) every application will be funded; it doesn't seem to happen that way. But if you work at it, you can significantly enhance the odds that your requests for grant support will be successful. By "work at it," I mean, first, seek guidance and instruction. Your mentors have valuable experience with this aspect of scientific life, and you should draw on that experience, seek their advice, and take advantage of their successes and failures. Their advice may not always correspond to the suggestions I'm offering you here, but that is exactly why you should seek other input. Everyone has a unique experience with grants, and your preparation will be enhanced by the disparate ideas that mentors and colleagues may offer for maximizing funding chances. In the same vein, as you begin to generate your own applications, ask colleagues—especially those with positive grant-getting track records—to read and comment on early drafts. Although you may feel embarrassed to ask for input on an "unfinished" product, the truth of the matter is that most scientists rarely feel that their grant application is finished—they simply submit it in the latest draft to make a deadline. Asking for help early in the process will save you valuable time and energy and may prevent you from going down a path that is unlikely to be fruitful.

Second, practice. Take every opportunity to write and submit grant applications. It may be painful at first, but it will definitely get easier with repetition (although the receipt of criticism may not). A very good way to learn about writing good, fundable grant applications is to participate in groups that actually carry out grant reviews. Membership in such review groups is generally invited and will depend on your having made a reputation for excellence in research in your chosen field. The time and effort spent in reviewing the grants of others often seems daunting, and many individuals are reluctant to participate in this grant-review process even when they are asked. But there are definitely rewards from this investment of time and energy. One of the principal benefits is gaining first-hand experience about what reviewers actually look for (and respond against) in a grant application.

As you gain experience and expertise, you will undoubtedly become more facile in this practice of grant-writing. That facility comes with a potential danger, however. We become all-too-accustomed to writing in "shorthand," using the jargon of our special field and making assumptions about what "everybody knows." While that usage may sometimes work in a manuscript to be published in a specialty journal, it rarely fares very well in grant applications. Using specialty jargon is an especially bad idea when applications are to be read by lay reviewers (e.g., nonscientists from private foundations that are providing the grant money). A good grant application should be intelligible and convincing to others with familiarity—although not necessarily technical expertise—in your field of interest. The members of the grant-review group that considers your application are likely to be smart, capable, experienced scientists, but it is unlikely that all of the group will be specialists in your field. Nevertheless, those individuals who are unfamiliar with your special area of research are still likely to be voting on funding decisions. Thus, it's important for you to make your message clear across a broad range of scientific interests and expertise. It's up to you to convince a reviewer outside your field that the topic of your proposal is significant, and that you have a research design that will result in interpretable results. Good science and a little salesmanship go a long way toward making this case.

◆ Summing Up

As a research scientist, you will spend a significant part of your career looking for support to fund your research. That search inevitably will involve writing grant applications. Depending on your position, these applications will be more or less elaborate, for large or small sums of money. In most cases, your application will compete against applications from other investigators who also think that their ideas are worthy of support from the granting agency. Because your research career depends, in large measure, on the success of these applications, this aspect of the research business can be very stressful.

It can, however, also be enjoyable. Good grant applications reflect your creativity and enthusiasm for your subject area. Writing them gives you an opportunity to think creatively, to exercise your command of your research area, to make the case for what you believe is important. In short, these applications push you, the researcher, to think carefully not only about a particular set of experimental manipulations, but also about how your work is likely to move us toward a significant goal. You will, after all, need to convince your reviewers (be they peers on a government review panel, officials of a private funding agency, or your bosses in a corporate setting) about the importance of your work.

**Real-Life Problem**: You submitted a grant application, and the review is negative. Your submission is the second attempt to get funding for this set of experiments, and the reviewers' scores are only marginally better than their scores for the initial submission. In going through the critiques, you're convinced that the reviewers simply didn't read your application carefully. What strategy should you use in resubmitting the application?

1. Write a strong letter to the review committee, pointing out their errors.
2. Rewrite the application, making an effort to accommodate all of the reviewers' suggestions.
3. Send the application to another granting agency.
4. Give up on that research direction and try something new.

**Discussion**: Each of these alternatives has something to offer. For many granting agencies, a resubmission should be accompanied by a clear description of how the application has been revised to address the critiques of the reviewers. In that context, you might craft a strong but respectful discussion of the points missed by the reviewers. Avoid accusing the review group of carelessness; rather, couch the response in words such as "perhaps I did not make this point sufficiently clear in my earlier submission." In general, if all you do in such a revision is to point out what earlier reviewers appear to have missed, you are unlikely to be successful. It is also necessary to make a serious attempt to rewrite your application with a focus, perhaps, on improving your experimental design, providing stronger preliminary data, and/or eliminating elements that were heavily criticized. In the case presented above, it may be that rather than simply not reading your application carefully, the review group simply didn't like your proposal (not interesting, not important). In such a case, the better strategy is to send the application to another granting agency, perhaps one with a mission that is more consistent with the goals of your application. If even that strategy does not bear fruit, then it's time to move on to a different research problem.

# 7 ◆
# The Politics of Science

◆ The Business of Science is not so Different from
   Other Professions

Although one hears occasionally about the romance of scientific research, there is really no such thing as the "ivory
tower" scientist—the individual who follows his or her
quest for knowledge and insight in isolation, with little
input (or interference) from the surrounding society. That
ideal may have been a reality a century or two ago, and perhaps the myth has been encouraged by the fact that so much
of research science is done within a university setting.

> But the fact of the matter is that "doing science" is
> a real-life occupation and entails all the problems
> that one encounters in any other life occupation.

Like every other job in our modern world, scientific
research involves interactions with people. Therefore, as a

scientist you will inevitably—unavoidably—be involved in what most people would call the "politics" of the profession. These political interactions consist, at their core, in trying to manipulate the system so that you achieve a place of prominence in your field. The term is applied to the give and take, the negotiation, the strategic positioning involved in getting what you want. Since in science most of your colleagues—who are also jockeying for priorities—are smart and ambitious, achieving such prominence is particularly difficult. It's rarely the case that you can outsmart everyone else, so don't enter the field if you think you can resolve all your issues simply because you're smarter than your colleagues. As is the case with any other job, you can try to minimize your involvement in these political interactions, but you can't make them go away. So a key aspect of the job of the scientist is to develop appropriate, socially acceptable strategies that will help you in your efforts.

Political interactions appear in almost every aspect of the research endeavor—or at least so it seems. There are politics associated with the process of grant application and funding. There are politics in the struggle for promotion and tenure. There are politics in the games played to gain more laboratory (or other) space. These manipulations may simply involve clear thinking and forceful presentation of your point of view. Often, however, they involve compromise.

> Politics, after all, consists of the practice (art?)
> of finding out how *everyone* can get what he or
> she wants.

From the point of view of a given scientist, the political process often means making it easy for those in power to grant your requests.

This chapter is not written as a lesson in scientific politics, but it is intended to make you aware of the importance of this aspect of the job of the research scientist. As such, I have one simple suggestion: don't become a scientist because you think you'll be isolated from the regular ("dirty"?) business of real life. The scientist faces the same types of challenges and problems, and has the potential of realizing the same types of rewards and benefits, as those in other professions. That's not to say that everyone in science must be a "politician"—at least not in the way we normally think of that term. But it helps to realize that you *will* face political problems along the way, and that these problems are a normal part of the scientist's life. Therefore, developing some political skills can be an important asset to your scientific life.

This chapter is designed to introduce you to some of the types of political issues that are now common in the business of a research scientist. The list is, of course, not comprehensive. And as in the preceding chapters, the viewpoints are my own—and I make no claim about correctness or universality. The discussion will, I hope, give you a taste of what political challenges you're likely to encounter as you develop a career in research. In my own view, an important aspect of "politics" is the business of integrating "how" with "why." As I've emphasized above, research science addresses the "how" question. But especially in the current social climate (and given the fact that much of our research is supported by others in our society—be it support from tax dollars, charitable donations, or corporate profits), we are often faced with the need to address the "why" question. This requirement is perhaps unfortunate, given that the "why" questions really can't be answered on the basis of research results—that is, if we keep on our scientist hats. But some attempt to integrate the "how" with the "why" has become a part of the job.

◆ Prestige

One can appropriately view the politics of science as a natural consequence of our striving for prestige, money, and power. Sound familiar? It's not unlike politics in other occupations/fields. We all do it, as individuals and as institutions. The one possible difference is that in science, the issue of prestige usually (not always) trumps money and power. So let's start with "prestige." What, exactly, does that mean in science? In a sense, it's one of those things that we all recognize but have difficulty defining. For an institution (e.g., a university or research institute), and even for a country, scientific prestige can be quantified by counting up such variables as number of highly competitive awards (e.g., Nobel Prizes), number of competitive grant awards and total grant income, number of investigators and laboratories, exclusivity (e.g., how hard it is to get a job there), and reputation of the faculty (frequency of publication, impact factor of the journals in which the faculty publishes). At the institutional level, prestige rests primarily on of the prestige of its members (e.g., the faculty at a university).

These variables also translate to the individual level, so that an individual scientist may "count" prestige based on number of awards, association with prestigious institutions, grant income, and so forth. These variables are obviously of subjective value, and so "prestige" will always be a subjective—and "relative"—measure. Nevertheless, in a given field, most participants have a similar set of values, and the highly valued factors (e.g., a Nobel Prize) will almost always confer significant prestige on the individual as well as on the institution. To be sure, the precise meaning of a given award or publication in a given journal will vary from field to field, but the ultimate goals are similar. There is constant jockeying at all levels—among individual scientists, across institutions, among countries—for recognition as holders of these highly valued (prestigious) factors. And within a field, there is an ongoing comparison in order to establish a given institution (or even an individual) within the relevant "pecking order."

Why should we work so hard to establish a prestigious reputation? Aside from the basic satisfaction associated with a reputation as one of the best, there is no question that this "reputation" translates into more tangible factors, such as money and power.

◆ Money

More and more in modern times, the major factor driving scientific politics is money. That condition arises from the perception—whether fact or not—that there's a limited pot of funding to divide among an ever-expanding group of "applicants." Thus, there's more and more intense competition for those things that we, as scientists, view as rewards, such as recognition, grant funding, and publication in the best journals. In theory, that situation would not necessarily spawn a "political" climate. If the criteria for evaluation were strictly objective, then the only issue of relevance in this competitive environment would be the quality of the science. But let's take, for example, the competition for grants. For most of us, we enter into a system of peer review. We put our wares in front of our peers and expect them to use objective criteria for judging and awarding grant funds. Although it's the best we have (indeed, there appear to be few alternatives that are seriously considered), it's an imperfect system. First, the objectivity of our peers is always questionable—they are, after all, part of the competition. If they are working in your field, they will undoubtedly feel that you are their competition, and they all therefore have a motive (conscious or unconscious) to reduce your competitive edge. The vast majority of grant reviewers do try to be fair, but they are human. Second, each reviewer has his or her own particular agenda with respect to what topic areas are worth supporting. So even if your science is impeccable, your choice of topic (including your goals and their significance) will generally meet with at least some skepticism. Third, the reviewers, expert though they may be, often disagree on the strength and value of a proposed set of experiments. So the fate of your project may rest on the question of how good a salesperson you are compared to your competitors. Added to all these variables are the programmatic pressures that arise from the priorities of the funding agency. Even a large agency (NSF, NIH) feels external (and often nonscientific) pressures (e.g., from Congress), and funding decisions may revolve around those pressures as well as the excellence of the science. All these factors ultimately lead to an opportunity for many nonscientific influences to seep into the decisions about what, and who, gets funded.

As a scientist, one might well ask what the justification might be for the involvement of nonscientific priorities in the support of the research enterprise. Why should your congressman, or corporate executive, or the lay public have so much influence on what we do? One answer, of course, is simply that they can—they have the resources. But there is a more important answer that is often neglected by those in the laboratory.

> What we do in the laboratory has implications—
> often profound ones—for the society around us.

Social priorities must be acknowledged as the major factor that drives the direction of laboratory research. We don't do science in a vacuum, and those who support the enterprise will, naturally, want to set the priorities.

Thus, as one enters this field, it's important to realize that everyone has a personal view about what constitutes good science, what types of research and which goals are most valuable, where the funding agencies should invest their resources. Given the differences of opinion that exist on these issues, the division of resources (e.g., grant funds) will certainly go counter to the views of at least some segment of the competing population. Your grant may not be funded. The review panel may not like your ideas. There is an unfortunate tendency—but certainly understandable—to consider such decisions (that go counter to your own views) as "political." As used in this way, the term certainly has a pejorative association, as a way of expressing disapproval (or moral outrage). What the term "political" really conveys in this context, however, is that the decision-making process is not "fair" and involves something other than objective scientific criteria. When you hear yourself speaking in these "political" terms, it's time to step back and consider a couple of the realities of scientific research: (1) reasonable (and smart) people can legitimately disagree and (2) evaluation of scientific research is not an exact or strictly objective process. Strange but true, scientists may not act in a strictly logical, consistent, and objective way when it comes to competition for professional rewards.

The involvement of nonscientific criteria in the division of the resource pot is, for better or worse, a fact of the modern research field. Indeed, it's difficult to imagine how it could be otherwise. Nevertheless, it's hard not to be aggravated by the perception that someone who knows a lot less than you about your area of expertise has the ability (power) to make critical decisions that affect your activities. We all end up complaining, at one time or another, about the apparent truism that it's who you know (your connections), not what you know (not the value or excellence of your work), that determines how much of the resource pot you'll get. Somehow, one tends to hear this complaint primarily from those who don't get what they want (or think they deserve). Of course, it really does matter who you know and what kinds of personal connections you have. Decisions (e.g., for grants, resources, promotions) are also based, to some extent, on other nonscientific variables (e.g., the prestige of your home institution, the prominence of your laboratory); these are factors that may provide useful information about how well you were trained, how successful you've been in past efforts, how you are regarded by your peers. So inevitably, political pressures (i.e., something other than scientific merit) will play a role in determining who gets what. There is no such thing as dispassionate, objective science (indeed, one could argue that scientists are particularly passionate).

Everyone has a personal ("nonscientific") agenda.

Your job, in part, is to make your agenda attractive to the larger world, including those who have the resources you seek.

In these struggles for money and other resources, it is important to remember that the availability of funding is cyclical. And while this feature certainly is true of grant funds, it can also be seen in availability of jobs, the abundance (or lack) of laboratory space, and so forth. What is striking in this cyclical picture is that when resources are abundant—when grant funds are easy to obtain—there appears to be less "politics" involved in the research process. The cyclical nature of support for research is in some ways encouraging; it means that if times are tough, they are likely to get better. However, it is also the case that funding decisions made at a particular phase of the cycle (e.g., in times of plenty, to expand large research programs) may have unintended consequences in the long term; for example, commitment of large sums of money to these large programs means that there will be little left for "the little guy" when funding becomes tight.

Funding for science, as for other social needs, is rarely part of a long-range plan to keep research laboratories vigorous and productive. Funding decisions are often short-sighted, based on immediate priorities. As such, these decisions can endanger the continuity of our research effort, particularly in making it difficult for young scientists to enter the field. Although we have learned enough to provide special opportunities for young investigators to obtain support, a very difficult funding environment may ultimately discourage a generation of trainees and leave us wondering where the research leaders of the future will come from.

◆ Power

In the above paragraphs, I've concentrated on tangible resources (e.g., money, lab space) as the basis for the political struggles so common in all professions. But as in other occupations, there is almost always another major factor driving the politics—power. The term "power" is also often used pejoratively, but in research science (as in other professions) it certainly has positive as well as negative potential. It is often intimately linked with "money," and there's little question that those who control the purse-strings also wield the power. But power has its own attractions and should be considered as a separate influence. Indeed, there are at least two distinct forms of power that are worth noting. Power, as influence over other people, will be discussed at length below. But as a scientist, perhaps the most important form of "power" is the ability to control your own working environment. That feature of research science cannot be overestimated and is certainly central in recruiting many bright minds to this enterprise. Under normal conditions, research scientists actually have an incredible amount of freedom to determine how they spend their time and on what projects they will focus their energies. That is real power. However, in the absence of supportive resources, that type of power may erode quite rapidly.

In the environment of scientific research, there appears to be no such thing as "absolute" power; all power is relative. For example, the power of the head of a rich research foundation is limited by many factors, including the expectations of both the funders and the lay public. Power in corporate laboratories is dictated largely by the success of the research enterprise, and so the corporate CEO is in some respects a slave to the abilities, insights, and energy of those who work in the laboratory. For large government funding agencies like NIH and NSF, the institute or foundation director's power is limited to encouraging particular granting mechanisms and establishing broad research priorities; the actual funneling of funds rests, for the most part, in the hands of the review committees (i.e., members of the research community). Nonetheless, those in leadership positions (i.e., positions of power) can make a significant difference. They are often instrumental in shaping the direction of research in a given field. This influence is the feature that attracts many of the best scientists to these often-thankless administrative positions. Exercise of this type of power often reflects a real desire by the individual at the top to make significant contributions to a field and to society by furthering the research directions that he or she believes are important and are likely to have significant payoffs. Thus, while those of us with relatively little power often use the term "power" pejoratively, it's unfair to cast such influence in an entirely negative light.

The exercise of power and influence is another general fact of life. Research science, as other professions, is often not an equitable business. So it does pay, as painful as it might sometimes feel, to develop relationships with those in power. You might call this effort "currying favor." It may well be, but it is also a politically smart and realistic move. Developing a politically astute mindset doesn't need to compromise your interests or ideals and can be beneficial when you need some help.

It is important to add in this discussion of power, as in the above paragraphs on money, that many science policy decisions are made by nonscientists, on the basis of nonscientific criteria. You, as a scientist, might be able to separate your science from politics (and from social, ethical, or religious issues), but you can't ignore these other influences— and they will certainly not ignore you. Given that the money used to support your research comes from the society around you—tax dollars, donations to nonprofit foundations, and even corporate income (i.e., the purchasing dollars of the public)—it's hard to argue with the "right" of the public (e.g., politicians, nonscientist foundation officers) to intrude upon the scientific enterprise. Especially in an environment of shrinking resources and budgets, this intrusion is likely to become more and more of a force in determining research directions. Rather than fight against this powerful stream, it may make sense to ask how you, as a scientist, can best contribute to the determination of these (often nonscientific) priorities. That becomes your job as a citizen as well as an advocate for the research laboratory.

Who sets the priorities, and who gets the resources? As mentioned in the paragraphs above, setting research priorities is not always (indeed, is rarely) a strictly "scientific" function. Priorities are established by people in power, often based on political pressures. A given scientist may disagree with those priorities; such disagreement usually derives from the perception that those priorities are not consistent with that individual's research direction (and therefore will mean reduced resources for that scientist's research goals). But there is a legitimate basis for setting "programmatic" priorities, for investing in socially important goals. Few scientists would argue against this practice, only against the choices. There are, however, serious arguments against various methods of resource distribution. For example, there is a growing trend toward funding large laboratories with significant track records. That tendency makes some sense from the point of view of the funding agency; the choice is "conservative" in the sense that one can invest in such a laboratory with confidence that the proposed work will be done. However, this allocation of resources does feed into the "the rich get richer" scenario. Some administrators will argue that this distribution of resources results in a "trickle-down" effect (e.g., the large lab establishes collaborations with smaller labs and/or employs large numbers of researchers). However, the actual result is a squeeze on small labs; in particular, it becomes increasingly difficult for young investigators who are just starting their research programs. It may also result in research that smacks of "self-fulfilling prophecy"—that is, studies that simply show what the major labs already know (or think they know). As a result of this strategy, imaginative new directions are often discouraged (or at least not encouraged), with the maverick and dissenter receiving scant attention. Funding large operations, at the expense of multiple small labs, certainly results in a gradual reduction in the number of "inputs" to a given problem. If one believes that it is important to encourage contributions from multiple sources, this method of resource distribution can have serious long-term negative consequences.

As indicated above, this problem is less acute when resources are plentiful, when there is enough for everyone. But when resources are scarce, those in power are faced with decisions about not only what problems to support, but also how to distribute the limited resources available.

It takes outstanding leadership to navigate through the dangers of this highly charged "political" environment.

### ◆ Leadership

Success in scientific politics often depends on your leadership qualities. But what constitutes a good leader in the context of research science? These qualities are not so different from those we'd like to see in any leader—say the president of a country. A leader must have long-range vision and effective administrative skills. Accomplishing difficult goals usually requires negotiation and compromise, in a way that results in all parties benefiting from the interaction. To convince others about the value of his or her vision, a leader must be respected and trusted. In science, these latter values usually result from a record of excellence in research, as well as a history of interaction with colleagues that reflects high ethical standards. Leadership style (e.g., dramatic or quiet, out-front or behind the scenes) is relatively unimportant, at least compared to leadership substance.

How does one become a leader? In science, the first step is invariably establishing a high-quality research record, a record that is admired and respected by your colleagues. This process, and the attendant development of those special skills involved in leadership positions (e.g., organizational, goal-setting), is typically a result of an implicit apprenticeship process. There are, of course, opportunities for explicit leadership instruction (indeed, there are courses and seminars that claim to prepare individuals for such roles). But in reality, few young scientists start off with the goal of becoming chairman or director; most just want to do good research. The pressure to assume a leadership role is often subtle, perhaps even a bit insidious at first.

But everyday exercise of the research enterprise inevitably starts one on this road. For example, every scientist must set priorities for his or her laboratory and establish long-term goals. The lab leader mentors aspiring trainees, thus directing the activities of other individuals. Almost by default, therefore, the scientist is asked to assume leadership roles.

Given this "automatic" involvement of the research scientist in politics (i.e., as a leader—or as a follower), what can an average young scientist do that is politically advantageous?

- Establish a reputation as an authority in your field. That is, do good work, publish in premier journals, and talk about your work at major conferences.
- Seek positions of responsibility and influence. Such positions might include leadership positions within research structures (e.g., the PI on a grant) but also in educational positions (e.g., head of a graduate/training program or training grant). Many individuals will sooner or later aspire to major academic positions (department chairs, deanships, institute heads), but such positions normally assume that one is already well established and well respected.
- Establish collaborative research activities. This activity is something that even the most junior of investigators can do. It will not only introduce you to others in the field but will also begin to challenge your organizational and interpersonal skills. At the very least, establish cordial, respectful, professional relationships with your colleagues—peers, mentors, bosses, and (yes) students.
- Join and actively participate in professional organizations. Become involved in committee work and participate in (and help organize) conferences, workshops, symposia.

- Become known to journal editors as a willing and insightful reviewer, and seek/accept positions on editorial boards.
- Look for opportunities to participate in grant-review committees.
- Initiate and/or collaborate in editing published works (books and journals). Or, if you have a lot of time on your hands, write the authoritative review work in your field.
- Participate in—or better yet lead—multidisciplinary, inter-institutional activities. Leadership of center grants, program projects, and collaborative research programs usually assumes you've already attained a position of respected leadership. It might pay to apprentice yourself to the leader of such a program and learn the ropes.

Of course, you don't have to engage in an active way in any of this type of "political" work to do good science. The need for political engagement will depend a lot on the type of job you choose and the targets of your ambition. A scientist who runs his or her own laboratory will inevitably become a politician in order to survive—even in relatively small and unpressured environments. A research fellow, on the other hand, working under the leadership of a "boss," may not have to deal with many of these political inputs. Your job choice, the choice of laboratory type, and the prestige of the institution will all contribute to the degree of political activity that is required of you. But whatever that choice, it is important to realize—from the outset—that science is a social activity. It involves meeting and working with other people (e.g., making contacts, networking). And this activity is a pathway to—a prerequisite for—influence and power. It's also the case that these "political" aspects of science can be fun, interesting, and challenging—as well as useful and rewarding. Being a skillful politician often goes hand-in-glove with making significant contributions to your field and to society. Major contributions almost always involve the joint efforts of many investigators, and productive interactions within such groups require skillful management.

The politics of science requires a somewhat different set of skills from those needed to do good laboratory research, and it's certainly true that one can be good at one and not be so skillful at the other. They do, however, complement each other—and those who are good at both can go a long way.

---

**Real-Life Problem**: Your lab is crowded and you need more space. Your department has some unoccupied space nearby, but there are several other faculty members who would like that space. How do you go about obtaining that space for your own lab?

1. Move some of your equipment into the space so that you are the de facto occupant (squatter's rights).
2. Make your case to the department chair.
3. Work out a sharing arrangement with other faculty who also want that space.

---

**Discussion**: This type of problem is quite common, especially given that we all (or almost all) think we need/deserve more lab space. The first option is a favored strategy that sometimes—but not always—works. The problem is that you can be easily kicked out of the space in which you squat if your chairman decides that the space should be used for other purposes. In such a case, you have no recourse and will have wasted considerable time and effort in setting up your operation in that new space. Further, "squatting" can trigger significant irritation in your colleagues who also want the space. It's almost always better to go through official channels so that you have clear authority for your move. Making a case to your chairman should, however, be developed with a compelling rationale, detailing not only why you need and deserve the space, but also why assigning the space to you will be of general benefit to the department. Perhaps an even better approach is to work out a sharing agreement (or trade) with other faculty. With the cooperation of several investigators, the group can make a case to the chairman that is likely to be much stronger than the request of a single individual. This latter approach also carries with it a potential "win–win" outcome in which everyone is happy. Especially if you're the one who organized such an effort, you will likely gain additional respect and support from your colleagues in the department.

# 8 ◆

# Ethical Conduct of Research

◆ Responsible Conduct of Research

After years of neglect, there is now a growing awareness of, and focus on, ethical issues involved in research. There are increasing pressures to develop rules that regulate our conduct in the laboratory—that is, how we carry out our experiments, how we analyze our data, and how we report the results of our studies. Such rules and guidelines have become particularly important as the public distrust of research science has increased. These pressures come both from within the scientific community and from the non-scientific public. There are now departments within virtually every research institution that provide education and training on various aspects of responsible conduct of research (RCR). There are organizations and associations that provide guidelines for publication ethics and that influence the way we use animals and human subjects in our research. These, and other RCR-related committees and agencies, have introduced a significant number of "hoops" through which the researcher must jump. (If you're particularly interested in aspects of this topic, you might want to participate on one or more of these committees, and contribute your energy toward making the business of science safer and more trustworthy.)

Depending on your field, the details of research ethics may differ somewhat, but there are certainly topics and general principles that cut across disciplinary boundaries and affect virtually all researchers. These principles include:

- Truthfulness in analyzing and reporting data
- Acknowledgement of (and giving credit to) those who contribute to the research
- Transparency in identifying the sources of research funding/support
- Behavior consistent with membership in a synergistic community of colleagues
- Establishment of goals that generally advance knowledge and contribute to the public good

These points seem rather obvious and are probably not worth belaboring, but they do play out differently depending on what type of institution employs you, what type of laboratory you work in, and what types of problems you're interested in. For example, the restrictions regulating the sharing of research results (see below) may be quite different depending on whether you work for a government laboratory engaged in top-secret research, a university laboratory, or a laboratory associated with a for-profit corporate enterprise.

Within these general areas of concern, there are specific issues related to:

- *Experiments on people.* Protocols for any study involving human subjects must be approved—*before* the experiment is started—by the relevant institutional review board. There are now widely accepted regulations that require (among other things) informed consent by the subject, protection of the subject's identity, and efforts to minimize any potential discomfort, pain, or negative health-related consequences of the study.

- *Experiments on vertebrate (and particular mammalian) animals (including nonhuman primates).* Institutional animal care committees similarly regulate experimental procedures on animals. Experimenters are asked to explain how they will minimize discomfort or pain, to provide a rationale for using the specific animal species proposed, to detail how they've determined the number of subjects they need for the experiment (with a goal of using the minimum number that will provide statistically significant data), and so forth.

- *Experimental design.* Guidelines now exist for developing an experimental design that yields interpretable results. Concerns revolve around such issues as the inclusion of appropriate control groups, nonbiased methods of data collection and analysis, and the use of sufficiently large sample sizes to guarantee statistical significance when appropriate statistical tests are used.

- *Publication ethics.* Journals (and publishers) are now extremely sensitive to any hints of inappropriate behaviors in publication, such as plagiarism (quoting work from another author without giving the appropriate citation), authorship (see Chapter 4), and duplication of material (multiple publications repeating the same results of an experiment). Concerns about publication ethics have led to the formation of international committees (composed of journal editors and ethicists) that advise journals about how to deal with difficult publication issues.

- *Conflicts of interest.* Similarly, there has been increasing attention to the potential of commercial (financial) influence on experimental design, analysis, and reporting (e.g., biased reporting, misrepresentation of data, suppression of negative results). Journals now routinely require authors to disclose their associations that might influence an unbiased approach to their study (e.g., a grant from a for-profit company that has an interest in the results of the study; stock in a company that might benefit from the results of the study). Similarly, professional societies that receive corporate funding to support their meetings and educational efforts must disclose that support (and require that their members do so as well).

• *Sharing of data with others in the field.* Especially given the potential for turning scientific (laboratory) findings into commercially profitable ventures, there has been an increasing focus on maintaining confidentiality with respect to some types of studies. While this situation remains the exception rather than the rule, there are some research environments in which investigators are not allowed to speak publicly about their studies. For example, pharmaceutical companies may require their own laboratory researchers, and scientists from laboratories with which they're collaborating, to treat the data as strictly proprietary. Similarly, results from some government laboratories that could have national security implications are routinely kept secret. Even among academic laboratories, the high level of competition sometimes leads to a reluctance of a laboratory to share its results with others, at least until those results are published and the laboratory's primacy in making a given discovery is established. As a result of these types of efforts, there is a renewed focus on the general issue of collegiality and collaboration in research science.

• *Environmental impact.* As in many other walks of life, there is new awareness of activities in the laboratory that may generate potentially toxic and/or dangerous materials. Guidelines have been developed to minimize toxic wastes, to develop appropriate methods of disposal, and to protect laboratory personnel (and the public) from harmful effects of these substances. Various regulatory agencies now routinely inspect laboratories to be sure that individuals are properly trained to handle and dispose of laboratory materials; violators may be punished with significant monetary fines.

• *Responses to various political, financial, or personal pressures.* While we would all like to think that scientific research is a strictly "objective" job, the fact that it is carried out by people means that there will always be nonscientific pressures that influence this enterprise. Efforts to minimize these influences (as outlined above) are helpful but certainly will not eliminate them completely. Our best protection rests in awareness of these pressures and the dangers they pose.

◆ Scientific Research as a Communal Effort

Why are these issues of responsible conduct so important to us? I would like to make the argument that in today's society, a scientist must be someone who is interested not only in the advancement of his or her own work, but also in the interests of the greater society of researchers. The idea of the scientist as a lone thinker—a Newton sitting under a tree, an Einstein laboring alone at his desk in the patent office—is no longer relevant (if it ever was, since even these apparently independent giants did not work in isolation). Today, it is clear from the outset of our research experience that the activity is an interactive one, involving a community of individuals who support and build on each other's insights and accomplishments. In such a system, what enhances and advances the careers of your colleagues will also be advantageous for you as an individual.

> Such a system is based on *trust,* a trust that derives
> from the belief that one's colleagues will play
> according to the rules that guide responsible
> conduct of research.

For example, we must trust that the data that are reported in the literature accurately reflect the results of the associated experiments; one investigator cannot, after all, replicate each study relevant to his or her area of research. Investigators who violate that trust jeopardize the work of their colleagues as well as our ability to move forward with new discoveries and insights.

Maintaining this communal perspective is sometimes a tricky business, since individual and communal interests do not always appear to coincide. This conflict can be seen in a number of different features of the business of research.

First and foremost, there is no question that research is a highly competitive enterprise. We end up competing not only with the anonymous "community," but also with our colleagues and close friends, for what often appears to be a limited set of resources. This competition is clear when our grants go to a funding review group that compares different proposals and prioritizes them for support. While we'd like to cheer on our colleagues, somewhere inside of us we are always aware that our chances for funding increase if our competitors do less well. Similarly, that tension is often expressed as competition between different fields or specialties. For example, if the total funding budget at NIH is fixed at a given level, increased awards for cancer research may be perceived to threaten the level of funding for Alzheimer's research. As a result, each specialty area may lobby for its own researchers—despite the general experience that a common lobbying effort (e.g., for a general increase in NIH research funding) tends to be much more advantageous (for everyone) in the long run. One can also see how communal cooperation (vs. the competitive instinct) might work advantageously at the university level; for example, efforts of individual investigators to obtain more laboratory space may be best served by communal support for a new research building that would serve and benefit large numbers of researchers.

It is the communal nature of our business that provides the basis for scientific advancement. We do not work in isolation. And as technical sophistication and the general body of knowledge grow in a given field, it has become almost impossible for any given individual or laboratory to encompass all that is needed to carry out cutting-edge research. So collaboration has become the name of the game. This requirement is now recognized by granting agencies, which are more and more inclined to fund research proposals that reflect the collaborative work of experts who contribute particular areas of expertise to the project.

The conflicts that arise as one tries to advance one's own career are also found in efforts to "get credit" for significant findings. Being the first one to report a significant finding (i.e., as the "discoverer") certainly carries rewards—publicity, prestige, patents, and so forth—so credit for this primacy is not an insignificant issue. In the world of modern research, it is hard to be unselfish, satisfied simply with the internal knowledge that one has made a significant contribution. Those who shout loudest and have the best contacts may get the credit, even if their claims of priority are not warranted. This issue of priority encourages a tendency toward secrecy and has spawned an incredible new movement in basic research—efforts to patent discoveries (e.g., genes) so that others cannot use them. The issue of how to regulate "intellectual ownership" is beyond the scope of this book (but see the section on intellectual property in Chapter 12) but is proving to be one of the most important and most challenging issues in our research society.

If you consider that these pressures—competition for resources and for credit—act on a group of smart and ambitious individuals (research science does tend to attract ambitious people with large egos), then it should be no surprise that the business of research can spawn unethical behavior. In fact, what is surprising is how little of it there seems to be! But it is, nevertheless, something that we cannot ignore, either in ourselves or in our colleagues. Responsible conduct of research means that we control these pressures and monitor our behaviors. For just as advances and awards for one individual reflect on the community, so too does negative behavior: publicity arising from the questionable activities of a single lab may have important consequences for all of us. If the public gets the impression that research scientists are willing to fabricate data or misrepresent results to further their careers, the public's willingness to support research will erode. That concern about unethical conduct is also a growing issue for those who control research resources (e.g., funding agencies, university officials).

◆  Engagement in Society

Responsible conduct of research also means engagement in society. Not only must scientists interact in their research community, so too must they play a role in the larger society. Increasingly, social debates require expert input. And you, as a research scientist, will often have knowledge that is appropriate for a given discussion. While most of us have traditionally been reluctant to get involved in political or social discourse, that decision is no longer a real option. It is sometimes important for you to make it clear where you—as a scientist—stand in the debate. To be sure, scientists often do disagree. These differences of opinion are evident at the level of interpreting the results of individual experiments, leading to the debate and exchange for which research science is justly known. When it comes to offering evidence and interpreting data within the context of social and political issues, these differences of opinion can be even more varied. Particularly since science does not engage in discussions about "truth," there is rarely a single scientific position on a given issue. Nevertheless, when the scientific community does come together behind a given issue (e.g., global warming), it can speak with a powerful voice that is respected and carries weight. Importantly, the respect and weight of the scientific viewpoint depend in large measure on the public's perception of science—and scientists. Scientific breaches of ethical conduct, real and perceived, can significantly reduce the weight of scientific opinion within the context of a more general public debate. Therefore, it is important not only to make your voice heard, but also to work toward developing and maintaining respect for the scientific viewpoint.

What kind of influence can the research community have on social and political policy? It is striking that our scientific discoveries (e.g., the rapid advances in medical technology) far outstrip the preparedness of social (and political) policymakers to make the most effective use of those advances. Scientists must therefore play a role in educating the lay community (including politicians) and in discussing how scientific breakthroughs can be implemented. There are many examples of the tension generated at the interface between new insights from research laboratories and uncertainty about what these discoveries should mean (or how they might be used) for social or political advancement: the use of DNA evidence, the "discovery" of global warming, genetically altered crops, the engineering of vaccines and other medical treatments. These advances all have tremendous implications—positive and negative—for our society.

It is certainly easy to take the position that many scientists have maintained in the past—that we're responsible only for making the discoveries, but not for how they're used. I am arguing here that in the modern world, researchers also should be involved in identifying the advantages and dangers of these discoveries and should help provide a balanced assessment of how they may be used. As a citizen and a participant in social progress, the scientist is entitled to an opinion like any other citizen. As a scientist, however, there are additional responsibilities for fair and reasoned argument based on real data. Depending on your field, you do have responsibility for discussing and arguing the implications of discoveries as they might relate to the environment, to health concerns, to military power and national security, and to the quality of life of the average citizen. These are heavy responsibilities, and not something that the average scientist thinks about every day. But as one considers a career in research, these responsibilities require serious consideration.

**Real-Life Problem**: You have been working on a research project, and it's time to write up the experiment and submit it for publication. As you've analyzed the study results, you've found the support of your hypothesis to be not quite statistically significant. How should you present and interpret these results?

1. Simply say that the results are not significant.
2. Use different statistics to try to squeeze significance out of the data.
3. Do more experiments to increase your sample size.
4. Point out the trend, and discuss the reasons that significance was not attained.

**Discussion**: Interestingly, all four of these alternatives are widely used. If the study has been well designed (i.e., with a sufficient sample size to adequately power the statistical analysis), the most straightforward interpretation of the appropriate statistical test is that the difference between/among experimental groups is not significant. But let's say your sample number is low, or the sample is not normally distributed. You might then see a need to do more experiments (i.e., increase the sample size) or to find a more appropriate statistical test to deal with your sample distribution. If neither of these options is feasible or helpful, you might want to simply discuss the trend, making it clear to the reader that the difference between groups does not reach statistical significance.

**Real-Life Problem**: You have published a study that suggests the feasibility of implanting laboratory-engineered brain cells into diseased brain tissue. A radio station contacts you to ask for an on-air interview to discuss the medical implications of your study. How should you handle this request?

*Continued*

1. Refuse the request, given the likelihood that whatever you say will be misunderstood and/or distorted.
2. Agree to the interview, but bring an attorney with you.
3. Agree to an interview, but insist that it be taped and that you will have power to approve—or not—the edited version of the interview that is to be aired.
4. Participate in the interview, but speak only on the findings themselves, without speculating about their eventual applications.

**Discussion**: This "problem" is becoming increasingly common. As I've indicated above, I believe that we as scientists no longer have the choice of opting out of speaking about our results (choice #1 above). And I'm not sure that bringing an attorney to the interview would be useful (choice #2); doing so would give the interviewer—and the listening audience—a rather negative impression. Asking for the right to listen to a recording of the interview before it is aired—in order to correct any misperceptions that arise from the station's editing process—is not unreasonable and would probably be acceptable to many radio stations. However, if the interview is to be aired "live," this alternative obviously would not work. In that case, if you agree to the interview, you will need to be clear about what you say as a scientist—as opposed to what you say as a citizen concerned about health care. Just as you would separate opinion from data in a research report, it would be important to be clear about that distinction when you discuss experimental results before a lay audience.

# 9 ◆
# Scientific Research as a Creative Enterprise

◆ Scientists as Artists

In our society, there is a tendency to contrast the scientist and the artist—the one working strictly objectively (according to rules of logic), the other calling upon a more free spirit. To a certain degree, these caricatures are appropriate, but it is important to realize that it is never "all-or-none"—not for the scientist and not for the artist. And it is even more important to emphasize that science—especially high-quality research—is a creative enterprise. One can "do science" by following the rules, keeping on the blinders, and trudging forward, but that version of research is rather boring, not only for those who read about the results but also for those engaged in the process. Scientific research is not always strictly "scientific," at least as we generally use this term. Research involves rules, but it also incorporates hunches, intuition, and serendipity. Indeed, without these latter components, it's unclear if or how many of the most important discoveries in science would have been made—from Newton and Pasteur to Einstein and Crick.

> I like to think about research as
> "controlled creativity."

Those who do it well are, in my view, real artists. They have brought this enterprise to an art form that combines true creativity with the discipline derived from intensive training. This description could just as aptly be applied to a painter or a writer—"disciplines" that we think of as creative and free but that also combine creativity and discipline in significant measures. To be an outstanding scientist or an outstanding artist, one must be flexible and open to new ideas. Changes in direction, new influences from one's colleagues, and "inspiration and perspiration" must all be incorporated into the work.

As suggested above, research science has many parallels with artistic creation. Here are just a few of them:

1. For both scientists and artists, "de novo" creation is a myth: all modern work is built on the insights and discoveries of previous generations. While there are occasionally quantum leaps, with radically new concepts and approaches appearing apparently out of the blue, close analysis always reveals important underpinnings from the past. It may not be strictly true that there is "nothing new under the sun," but it is almost certainly the case that what is new appears within the context of years of exploration and experimentation. Most scientists (and artists) are very much aware of their debt to previous workers and clearly acknowledge their starting points. Putting new work into a historical context is a significant part of any research report.

2. Although modern work builds on the development of past accomplishment, good science and good art generally make their marks by offering a new way of looking at (often old) issues. New perspectives, as well as new techniques, are key components of both enterprises.

3. These new perspectives most often result from the scientist's (or artist's) ability to *integrate* current and past approaches. While small steps in the field may be taken by following relatively linear pathways, major advances—the step functions of discovery—are usually the result of an individual's ability to see the big picture and manipulate all its elements in a unique way. That is the basis of "creation" for both scientist and artist.

4. Carrying out this creative task effectively—in science and in art—requires technical expertise. For both enterprises, expression of new ideas and development of new techniques is usually based on mastery of the existing approaches to a problem. Although one may certainly have innate "talent," mastery of technical aspects of the enterprise requires study and practice. And mastery is learned often through an apprenticeship process. Only the rare genius might (occasionally) get away without such training.

5. With that mastery of concepts and techniques, the scientist or artist can then "create" according to his or her own individual style. Both research and art are basically expressive exercises, with the practitioner investing much of his or her own individuality in the activity. Certainly, one can be a highly competent technician without being a creative scientist. And it is possible to take small steps as a "copyist" (i.e., using tried-and-true styles and approaches). But all research—and certainly the work that generates major advances in a given field—reflects the style of the individual scientist. While some critics may maintain that style is not substance, I suggest that style constitutes a significant part of it.

6. Finally, both science and art come in different forms and are employed for different goals. Each research (or artistic) enterprise has a different starting point and will aim toward different consequences. Both scientific and artistic processes can be traditional or exploratory, conservative or high risk. The work can serve to shock the field or to confirm previously held views. It may be highly personal or incredibly expansive. Good work tends to express the individual style of the researcher, but also has potential for social relevance and impact.

Here's another aspect of the research enterprise that has overtones of artistic work: Research science can be—and should be—fun. Indeed, if you don't find enjoyment in the work, you've probably chosen the wrong profession. Certainly, one can do "safe" science, which involves doing the same manipulation (of ideas, of formulae, of materials) over and over again. Sounds boring? Not fulfilling? Not exciting or fun? This kind of work has attractions for some people, who find enjoyment in the very essence of the repetition, the challenge of gaining exceptional expertise, and/or the opportunity of using the process (or the product) to move to another level of being. Some art is like that, too. Other approaches to science offer the opportunity to take chances and try out new ideas, with the potential of making exciting breakthroughs. Again, science mimics art in this range of activities, with the opportunity for different types of researchers to find fulfillment in different types of laboratory work.

One of the striking aspects of both research science and expressive art is how the activity—the execution—changes with practice and experience. The researcher (and the artist) inevitably learns from the successes and failures of past attempts and uses those insights to make the next attempt "better." This description is, I suppose, appropriate for most activities in life, but it seems to be particularly apt in research and in art. In many ways, both activities are essentially interactive, with each repetition revealing something new and valuable. One of the important lessons from this repetition is the understanding (realization) that there are many ways to approach a given problem—whether it's a research/experimental problem in the laboratory or a problem of artistic expression. And for both art and science, there are many ways for working out the problem (trying to find the solution). One can simply sit and let ideas "incubate." At least for science, this approach is highly effective. Teachers frequently tell students who are stuck on a problem to "put it away" and come back to it later. It's actually quite amazing how, without "working at it," we can massage and manipulate difficult issues into manageable ideas and processes. Alternatively, one can try out different strategies and approaches, taking advantage of the kinetic "expertise" that comes from practice and repetition to test out variations on a theme. These alternatives are both ways of individual exploration and often involve an intense, inward focus. But there is another, quite different tack. Artists and researchers alike often take their problematic issues to their colleagues to obtain different input (as seen with fresh, "objective" eyes).

◆ Personal Rewards of Creative Enterprise

Finally, the research process—and product—should provide the same personal feedback as any creative artistic enterprise. Both process and end product can (and should) be:

- *Exciting.* A good experiment, a predicted (or unexpected) result, an insight into how things work can provide a real thrill to one who appreciates and understands the field.
- *Thought-provoking.* No experiment—and arguably no really significant piece of art—simply "sits there."

They're both starting points, provoking reactions, ideas, arguments, and commentary. As such, they both—if done well—tend to spawn imitators. The great artists have given rise to "schools," in which the major features of the founder are carried forward by lesser artists. The same may be true for science, where schools of thought and schools of experimental approach arise around major figures or significant studies.

- *Beautiful.* Beauty is, indeed, in the eyes of the beholder. For the art connoisseur as well as the dilettante, for one who has studied the process as well as for the casual observer, beauty is perhaps a "given" aspect of art. The same can be said of a well-constructed experiment, of a study that produces insights and explains some of the mysteries of our life. Even the lay observer can appreciate this beauty if the experiment is explained properly.

- *Fulfilling.* Both scientific and artistic activities give rise to a rather puzzling but characteristic dichotomy. The work is, on the one hand, incredibly fulfilling; on the other hand, it is always not quite satisfying. For every piece of art, for each experiment, the "artist" can almost always say, "I could have done it better." That is certainly the case in research. Not only can one always "do it better," but there is a sense that an experiment (like a painting?) is never quite finished. The scientist comes to a stopping point when some external or internal force tells him or her to stop, not that "it's done." But despite the absence of complete closure, the scientist (or artist) is usually (often) in the enviable position of saying/thinking, "I can't think of anything I'd rather be doing." That's not to say, of course, that one doesn't occasionally (or even often) have doubts. But if you don't feel that science is fulfilling, then you're likely in the wrong field. Doing research science is too hard and requires too much of a commitment for it not to carry with it the potential for significant fulfillment.

**Real-Life Problem**: You've submitted a grant application, and the review comes back with some critical comments about the "creative" and "speculative" nature of the proposal. How should you respond to those criticisms, and what steps should you take to "curb" your creativity?

1. See if you can generate more "acceptable" specific aims by taking a more conservative approach.
2. Address the issues regarding speculation, but try to maintain the creative aspects of the proposal.
3. Emphasize the experimental background and logical thought processes that underlie the creative aspects of your proposal.
4. Provide an example of how your creative approach might yield new insights that would not have been seen in a more conservative, by-the-book approach.

**Discussion**: There are at least two separate issues involved in this problem—one that speaks to your "grantsmanship" skills and the other that challenges your creative impulses. My view is that all grants should contain both conservative scientific elements and more creative (i.e., higher-risk) components. The latter, if it is presented with a thoughtful explanation of the rationale behind the proposal, should add interest and excitement to any grant proposal. Even a rather speculative proposal requires scientific justification—including the thought process and the evidence that led to the proposal, and the potential consequences of

*Continued*

the proposed experiments (for the field, for society). The trick, always, is to present something novel and intriguing with a strong scientific rationale. As is often the case for creative (artistic) enterprises, what appears to be exciting and novel to the author/maker may not seem so to others (e.g., your grant reviewers). So you should be prepared for rejection—and decide on alternative strategies (conservative revision, submission to other granting agencies, moving on to other ideas for grant proposals) should you receive such rejection.

# 10 ◆

# The Role of the Scientist in Society

◆ The Role of Science in Society

As we discussed in Chapter 8, there is no question that scientific research is a social enterprise. The idea of the lonely scientist in his isolated laboratory, dreaming up various schemes entirely based on notions arising in his head, is a fairy tale. It certainly does not pertain to the modern scientific world, and I doubt that it was ever a reality. The social aspects of scientific research take at least three different (but certainly interrelated) forms: the relationship of a scientist to his or her professional forbearers (i.e., his or her relationship with history); the interactions among colleagues; and the interactions between a scientist and the society in which he or she works. In previous chapters I discussed the importance of understanding and appreciating the history of one's field. Interpreting data from the laboratory requires the awareness of previous contributions and the place of one's own work within that framework.

We do, indeed, stand on the shoulders of those who have preceded us. In the following chapter (Chapter 11), I will discuss further the second set of relationships (again, issues that I've touched on previously)—how scientists interact with each other. The present chapter is meant to explore the scientist's relationship with society, and in particular to provide an introduction to the question of the scientist's role in society. Do we, as scientists, have a special set of responsibilities? Not too many years ago, this question was rarely asked, since few people understood the profound effects that scientific inquiry and discovery could have on their lives. Today, that situation is quite different.

◆ Prediction

Modern science is a system, developed over a surprisingly short period of time, for explaining our world. It provides us with tools that lead to the perception that we can master the mysteries of the universe—or at least make reasonably accurate predictions about how things work (and will work in the future). In that larger sense, the system of science is not so different from religions, or from superstitious beliefs—which also have in common this drive to understand and predict. Indeed, this basic function of science, to make predictions, is poorly understood by most individuals, especially those outside the science enterprise. An underlying assumption of modern science is that if we have knowledge of the variables that contribute to a given event or outcome, we can predict when that outcome will occur—whether it be a disease, a tsunami, or global warming. Similarly, if we can make reasonably accurate "cause–effect" predictions, we can design spacecraft that will travel to Mars, manufacture more effective weapons systems, or develop new energy systems. Because science is based on probability, not certainty, there is always a possibility of failure—even when we think we know (and have controlled) all the variables that contribute to the predicted outcome. Such failures are often seen by the public as evidence that the scientific enterprise is flawed, makes false claims, and cannot be trusted (see below). But in fact, the potential for failure is an intrinsic aspect of any predictive system.

Although this book is not the place for arguing for or against the view that science is simply a "refinement" of religious (or superstitious) belief systems, I do argue that science and religion address fundamentally different kinds of questions. As presented in earlier chapters, the goal of religion is to ask (and answer) the question "why?" Science, in contrast, has a much more mundane and practical goal in working toward an understanding of "how?" This fundamental practicality has arisen, in large part, in order to develop ideas and machines that will make our lives better. In other words, the goal of science is basically one of social "improvement." There is, of course, the drive for knowledge for its own sake—but historically, science has been driven by (and therefore supported) social concerns. And therein lies our social responsibility as scientists.

And at the same time, as we work toward the better life, it is important to keep in mind that the scientific method is not the only way to understand the universe. Indeed, a surprisingly small proportion of our population, even within advanced cultures of the developed world, understands and agrees with the methods of research science. There is, to be sure, considerable "lip service" to the idea that science is important in building a better future. However, for most people, other (and older) methods for organizing and understanding the world (e.g., based on religions, philosophies, or specific political systems) feel much more comfortable. Indeed, there has been much discussion about whether science is compatible with other belief systems. Again, this discussion is much too big for this volume. I do believe, however, that there is no intrinsic opposition between science and other systems. They are simply different, with different methods and different goals.

◆ Social Responsibility

Given the fact that most people, in their day-to-day lives, don't understand—and perhaps don't accept—a scientific approach to the world, it becomes particularly important for you, as a scientist, to help demonstrate how the scientific approach can be integrated into everyday life. This challenge has become surprisingly difficult. While the word of the scientist used to be widely respected, that no longer seems to be the case. Perhaps we have hurt our own cause by making exaggerated (unrealistic) claims, by taking on apparently self-serving attitudes (particularly if there are profits to be made), or simply by withdrawing from social discourse. To be sure, there is the occasional "practicing" scientist—with excellent communication skills— who has attempted to explain recent advances in his or her field to a broad audience. However, all too often, in place of solid scientific discussion, we now have popularizers of science who present a "watered-down" (and often simplistic) translation of difficult and complex scientific concepts. Certainly, there is a need for effective communication with the public. However, since we scientists (as a rule) communicate poorly with nonscientists and have trouble describing our concerns and convictions in terms that are easily understood, we have been largely replaced by those who are primarily communicators. There is a need for scientists to argue their points of view in the public arena. According to numerous polls that ask the public to indicate what professions they most respect, we have somehow lost respectability and credibility. And that loss has provided an opening to critics of "institutional" science. According to these critics, we are too "political," too much wedded to the "party line." To make things worse, many of these critics have legitimate scientific credentials and can be very convincing.

So how do we respond to these critics? First, we should be clear about the argument. One of the easiest ways to level criticism is to confuse the issue. As a scientist, you should have a real advantage here, for you have made a career of learning to think clearly. Be clear about what is being proposed on each side of an issue—and don't accept the premise of the critics if it doesn't make sense. Beware of situations in which criticism is directed against inaccurate or extreme statements, often implicit and rarely clearly articulated. If one can identify the inaccuracies and exaggerations behind such arguments, there may actually be no argument at all, or the issue may be one of how strong a conclusion can be drawn from the data.

Second, don't forget that you—the scientist—can never prove anything conclusively. You can support a hypothesis with observation, and you can argue carefully and logically. But proof—the kind of proof that seems to be "required" of the scientist in these criticisms—cannot be offered. Scientists don't pretend to have discovered the "truth" or to know definitively that a given position is "correct." All you can do is draw conclusions based on the best available data.

Nonscientists are uneasy with this position (as are some scientists): they prefer certitude. It's hard to convince someone that, given all the time and effort and money you've spent on studying a problem, all you can do is offer the best likelihood—or several of them. Multiple answers may be confusing and dissatisfying, but it may be the best that the scientific approach can offer (at a given point in time).

Third, be careful to present your views—results of your own studies or critiques of other work—in a reasoned manner. It's all too easy to sensationalize results that have a "sexy" ring. But it's irresponsible to leave the public with interpretations or expectations that go beyond the current state of knowledge.

Fourth, avoid the trap of confusing correlation with causation. We all want to know "why," but clear-cut causes are rarely available as explanations. Correlations are mostly what we deal with. That can be seen dramatically in the current arguments about global warming. Scientists point to correlations (e.g., between temperature increases and carbon emissions from burning fossil fuels), and critics counter that all that has been shown are correlations, that there is no direct cause–effect proof. The correlations are quite convincing within the context of scientific method—and of course, the experiments needed to demonstrate cause and effect with respect to global warming are impossible. Science can do just so much.

Finally, it's important to acknowledge legitimate concerns, and it is important to respect the views of nonscientists, no matter how "unscientific" they may be. Common sense is an incredibly valuable tool, and one doesn't need to be a scientist to exercise it.

> At the same time, science is not a democratic process. Everyone doesn't have an equal vote.

You, as a scientist, will ideally bring a unique critical background to issues of concern, with special knowledge that can be helpful in assessing the situation. You can offer an "expert" opinion and point to the data that support your position, but you can't force anyone to accept your position, to believe in the scientific solution.

◆  Why Don't People Trust Scientists?

Not so many years ago, the word of the scientific community carried considerable weight, and scientists were highly respected. That seems no longer to be the case. Indeed, there is a considerable distrust of both the scientific method and the "conclusions" that are relayed to the general public from the scientific community. Science has offered an opinion about virtually every aspect of everyday life, ranging from dietary advice to pronouncements on global warming, from discoveries of new medications and medical procedures to insights about the structure of the universe. The public has grown increasingly skeptical about these pronouncements, and you (as a scientist) will need to be aware of that skepticism and make some efforts to change it. Why should there be such distrust?

1. The public doesn't really understand what science is, what it does. There seems to be a broad perception that the scientist, when he or she speaks, will offer the truth. But that is not, as I've said, the job of science. Rather, science is in the business of making predictions, so that we can better understand and "control" our world. The scientist says, "If you do this, then that is likely to happen." And the business of science is to test these predictions, refine them, and apply them to different types of questions.

2. Scientists tend to be rather condescending when dealing with the public—as though they are privy to some secret truth that the public cannot possibly understand. Certainly, the scientist has a special perspective, based not only on years of study, but also on a particular way of addressing a problem. But arrogance is not a useful tool.

3. We are sometimes wrong. Of course, since we're always testing hypotheses, being "wrong" is not seen as a problem by the scientist. But if the public expects "truth,"

then being wrong does become problematic. In this context, the occasional incorrect prediction may aggravate the perception that scientists are "in the pocket" of some for-profit enterprise (e.g., oil companies, pharmaceutical companies). Thus, mistakes can be seen as intentional and based on greed.

4. Scientist talk a language all their own and have trouble communicating their insights in language that nonscientists understand. This problem with communication can be viewed as obfuscation—again intentional. And we do little to alter that perception.

5. There are multiple sources of information easily available (on the web) to everyone, and websites often offer dramatically different points of view. Website "experts" may deliver opinions based on various types of evidence, often without rigorous scientific method. The self-educated layperson can choose among what appear to be many equally valid options for authoritative information.

◆ The Future of Science in Our Society

Given the suspicion with which science is currently regarded, and the speed with which our modern world is changing, it is not unreasonable to wonder what will become of our discipline in the future. Will social and political pressures force scientific research "underground"? Unlikely, but there are likely to be major antiscientific consequences for our discipline if the current climate is not altered. Perhaps more likely is the possibility that the electronic/digital age will change the essential nature of research science. How might our patterns of behavior change in the face of new technologies? Remember that modern science is, indeed, modern—it has been with us only a few hundred years at best.

Will science lose its social essence? I've emphasized the basically social nature of research in the paragraphs above, but ancient "research" was a much more lonely occupation. Will we revert to this isolated-investigator mentality in the face of fierce competition and limited resources? Will research again become the province of the rich and privileged, of genius and exception—or will we continue to draw from a more "everyman" tradition that currently brings us the benefits of diversity of background and outlook?

Whatever it becomes, you—as a young investigator—will surely have an important influence on its direction, so you must be prepared to fight for your perspective. Not for right or wrong, proof or superstition—but for the importance of the scientific approach. And remember, there has always been (and will presumably always be) an essential tension between the scientific viewpoint and the pressures and inertia of economic and political interests. The public, and politicians, like answers and certainty. The scientist offers likelihood and probability. The public seeks "truth," while we offer only the best explanation given the available data. If you believe that the scientific approach offers our society an important tool, then you should be prepared to take part in the debate. You can play a critical role—as an author, a speaker, an educator—if you so choose to do so. But remember, your scientific background does not provide a shortcut to the truth. So while you fight for the scientific approach, make sure you separate your science from your opinions.

**Real-Life Problem:** You are a guest on a radio talk show, describing a new discovery in an area of your expertise. You provide your best, and most careful, scientific description of the discovery, but callers start to attack you for being misleading or untruthful. How should you respond?

*Continued*

1. Try to explain that science cannot guarantee the correctness of its conclusions, only provide the most likely answers given the available data.
2. Explain the issue again, using simple language.
3. Agree that there are other possible explanations.
4. Cut short the conversation and refuse to appear again in such a public forum.

**Discussion:** Given my repeated argument that the modern scientist cannot opt out of the social dialogue, the fourth possibility suggested here is not one that I would support, although it has in the past been a popular response for many scientists. The other three options may all be appropriate, and they are not mutually exclusive. The public needs to learn that the scientific method cannot provide certainty, and that the accuracy of scientific prediction changes as better technologies and more evidence become available. These points, as well as the methods that are used to generate scientific data, can and should be explained in clear and straightforward language that a nonscientist can understand. Within this same context, a scientist should always acknowledge the possibility of other "answers" while making clear the requirements of scientific hypotheses — i.e., that any potential answer (hypothesis) must be tested according to rigorous experimental rules.

# 11 ◆
# Personal Challenges

While there are certainly important intellectual and scientific issues to consider in thinking about a research science career, they are not the only—or even the most important—considerations. As you try to evaluate your suitability for a research career, here are a set of more personal factors to think about.

## ◆ Personal Interactions

As a researcher—in academics or in private industry, as a student or as a supervisor, wherever you are and whatever you do—you will enjoy the support and challenge of a community of individuals, all of whom have in common the goals of scientific research. And indeed, this uniting feature may be the only thing you have in common with your colleagues. The research community is composed of the same diversity of individuals as one might find in other types of careers. With some individuals you'll have a positive experience, with others a negative interaction. Some will become friends, and some will be individuals you'll want to avoid. While the diversity may not quite mirror that of the general society, you can expect a broad gamut of personalities.

Most of us have a number of features in common. For example, most people who go into research science are smart—some very smart. A subset of these folks will not only be smart but will also think of themselves as brilliant. An overlapping subset of these individuals will be intolerant of those whose IQs are lower than theirs, and will not suffer fools gladly. Another common feature is that almost all of these scientists are ambitious. In some, that ambition will be more obvious than in others. But it is useful to assume that all of your colleagues (e.g., students, fellows, professors, supervisors) are ambitious. That reality is not so different from many other professions.

But as important as these similarities are, there are important differences across the scientist population. For example, while all of us are "driven" to some extent, different scientists are driven by different factors. Each has his or her own unique set of motivations and goals. Some are driven by the desire to unravel the puzzle. Some are looking for fame (but rarely fortune). Some see research as a safe haven in the midst of a chaotic social maelstrom. Do not assume that everyone is motivated to the same extent, and by the same factors that motivate you. Further, it is useful to realize from the start that you will encounter a variety of styles and social skills—features quite independent of scientific/research capabilities. For example, some scientists are extroverts and enjoy social interaction. Others are more introverted and prefer to be left alone. Some are as slick as used-car salesmen, but others have trouble stringing two sentences together (at least spontaneously). The variation in social skills means that you cannot take anything for granted as you meet and interact with your colleagues.

Appreciating these differences is particularly useful in early years, when young people are looking for the "right" laboratory and the "right" mentor. Those choices must be based on your perceptions of research quality and scientific reputation. But don't underestimate the importance of your potential mentor's style and personality: the wrong match can make life miserable for all concerned. And be aware that just because one of your fellow students finds a particular professor agreeable, that doesn't mean you also will interact easily and amicably with him or her. It's important to give such an important relationship a "tryout" to see how you get along.

Even with an appropriate trial period, trainees sometimes end up in the "wrong" laboratory, with the wrong advisor. Individuals who find themselves in this position should spend some time considering what, exactly, constitutes this "match" problem. Is it that the trainee doesn't like the kind of work he or she has been assigned or is likely to do for a dissertation? Is the research focus of the laboratory uninteresting? Does the student object to the way in which the advisor treats him or her (and others in the lab)? Identifying the issue will help considerably in deciding on the most effective course of action, be it attempts to move to another laboratory (if there is a significant personality conflict with the initial advisor), to improve the situation in the offending laboratory (change assignments, reevaluate the proposed dissertation research topic), or simply to give up the idea of a graduate career in this area of study.

It is not just your relationships with mentors (and other senior investigators) that can be tricky. Indeed, collegial relationships among peers can be difficult. As mentioned above, there is likely to be competition among "friends," with jealousy and one-upmanship interjected into what might have been a perfectly enjoyable relationship. Maintaining friendly and effective relationships with a diverse group of competitive and ambitious individuals can be a challenge, one that requires considerable "people skills." Nowhere is this need more evident than in the management of a laboratory. In this microcosm, individuals at all levels of training and accomplishment are likely to interact. Relationships will inevitably reflect—to some extent—the hierarchy upon which the lab functions. As the lab's PI, you not only will be "in charge" (i.e., in a position to direct traffic, to assign work—to tell others what to do), but will also be in a position to hire and fire. You therefore will need to develop a capacity for identifying the best person for a job (or the best job for the individuals at hand), to treat individuals fairly, to deal with misconduct and unsatisfactory job performance—and do it all while maintaining an atmosphere that encourages allegiance and loyalty, dedication, and hard work. Finally, even if you do an exemplary job, not everyone will appreciate your work. Indeed, from the outside, a well-run lab appears to function "automatically"; people tend to pay attention to your skills as a manager only if/when something goes wrong (or someone is unhappy). Only you will fully appreciate the contributions you make to the smoothly operating laboratory environment.

◆ Self-Esteem/Self-Confidence

Many young scientists, at some point during their training, question whether they are actually smart enough and capable enough to make a successful career as a researcher. This "crisis of self-confidence" is natural enough, since you are likely to be surrounded by other bright young scientists-to-be—all of whom invariably seem smarter and more confident than you. Indeed, part of your job as a trainee, and an important part of your "education," is to evaluate your chances for success within the context of competition with other similarly motivated individuals. You will find yourself making comparisons with your lab-mates and with other students in your training program. These comparisons do, and should, extend well beyond your training institution—for example, as you begin to go to meetings and symposia and meet scientists from other universities and other countries. These congregations of scientists are often largely populated by young people early in their careers, not only looking to make good impressions on more senior investigators, but also seeking opportunities to gain an advantage over their peers. These meetings can be very exciting but also somewhat intimidating and stressful (even for more senior researchers).

It pays to remember that, for the most part, these hordes of apparently self-confident colleagues are just like you. They, too, are wondering if they're good enough. And they tend to overcompensate for any sense of insecurity with a bravado that serves the purpose of impressing and intimidating their peers. In facing this question of confidence, you therefore have two jobs to work through. One is establishing an objective (as much as possible) view of your own relative capabilities. And the other is sorting through your competitors' smoke-screens that are intended to give inflated impressions of their accomplishments. That is not to say, of course, that everyone is "equal," and that there are not a set of particularly gifted scientists out there. It does mean, however, that not everyone is a superstar, and that your chances (if you're not a superstar) are perhaps not as dismal as may first appear.

The process of developing confidence, of developing a realistic self-image, is difficult. From my perspective as an advisor, it is perhaps the most important help that I can offer a trainee. And for the young scientist, developing that self-confidence goes hand-in-hand with the job of selling yourself. It would be nice to think that your stellar qualities will be immediately and automatically recognized by your colleagues, but that is not so likely to occur. Just as it's up to you to generate outstanding work in the laboratory, so too is it up to you to convince the world (scientific or otherwise) that your work is important and worth its attention. Self-promotion is a part of the business of research. And just as in the "real world," that process of selling yourself can be done in a way that impresses rather than alienates. From the time you are a student applying to graduate school or a research laboratory, the job of making a good impression will be an important part of your career. Your effectiveness in this effort starts with your believing in yourself and being able to convey sincerity and excitement about your work, about your ideas, and about your goals. Excitement is infectious.

◆  Commitment to Career

Scientific research is not a glamorous occupation. It is sometimes challenging, and sometimes rewarding, but it is often a drudge. There are, as I've indicated earlier, certainly easier ways to make a living. Further, to do the job well, to be successful, requires a commitment that often seems excessive. What is required is not just an intellectual (and often emotional) commitment, but also a commitment of time and energy that may exclude (at least for some period of time) most other activities. Depending, of course, on the type of research you choose, a young trainee or faculty member could easily spend 16 hours a day, 7 days a week, in the laboratory. And that time is not all spent in exciting activity. The activities that accompany the actual lab work— writing grants and papers, researching and/or reading the literature, analyzing data—are laborious. Ideally there will also be a few "eureka" moments that provide excitement and reinforcement; those moments, together with your enjoyment of intellectual growth and technical accomplishment, need to be sufficiently rewarding to encourage your long hours of labor. It's worth noting, also, that laboratory work does not proceed according to a 9-to-5 schedule: weekends and evenings are always at risk. And so there is the inevitable question: Can I have a life outside the laboratory?

The answer to that question comes in several different forms. The good news is that although your time commitment is considerable, the way you schedule that time is generally up to you (at least when you get to the point of running your own laboratory). Research, while demanding, has a type of flexibility that is rare in other "grown-up" occupations. In addition, as you grow in your career and attain more senior status, you will have more control over how you schedule your time. Of course, by that time, you may be a confirmed "workaholic," with the expectation that you will devote 90 hours a week to your job. Thus, the answer to that question about "a life outside the lab" tends to be answered with a promise of time and leisure "later."

I should add here that you can actually pursue a career in research, with a 9-to-5 schedule, if (a) you are not particularly ambitious; (b) you are not an "A-type" personality to start with; and (c) you do not need to be "in charge" and can be happy in a support role. Success—as measured in advancement, more responsibility, and greater rewards—is not likely to be reached without "working your butt off."

So what's the point of investing all this work in research science? How are you likely to spend all that time? As a trainee, perhaps by definition, you spend your time and energy in learning how to be good at what you do. Practically, you devote much of your time attaining technical excellence. No matter how many books and articles you read, there is really no substitute for practice. That's true whether trying to gain facility with a particular research technique or trying to become a proficient grant-writer. Once the technical aspects of the job are second nature, you can focus on the much harder jobs—selecting the questions and problems you want to target in your research, figuring out how to design experiments to obtain useful and meaningful data, and understanding what to do with a set of data once you have it. Skill with these latter tasks comes gradually, as you develop a broad vision of your area of research (and presumably a more sophisticated understanding of your field). Because the refinement of these capabilities often requires decades of practice and experience, one can realistically expect that the commitment required in your job will be ongoing throughout your career.

◆ Discrimination and Diversity

As in many fields, and certainly in many "professional" fields, the hierarchy in research science has long been controlled by "old white guys." Discrimination against young people, nonwhite people, and women certainly exists, although one would like to think that these attitudes are much less influential now than in the past. Interestingly, forms of discrimination seem to differ dramatically across different types of research science. For example, in our modern society, some disciplines are now dominated by women. Women can and do rise to positions of leadership in various research fields and certainly have a solid history of significant contributions. Despite these advances, it is important for us to stay vigilant, to remain conscious of the traps of discrimination, and to work toward equity. Rather than ignore it or pretend it doesn't exist, it's worthwhile spending a little time exploring some of the issues that women and minorities may face in research science. While some of the topics I've introduced below are not "politically correct," I present them as food for thought—and as a challenge to all young scientists.

My own view is that whatever their capabilities, women in science usually face a more difficult set of challenges than men. That may not be because of overt discrimination, but simply because "the system" has been established by men, for men. Aside from the obvious traditional domination of some fields by men, there is the general difficulty of trying to balance family and career. As I've indicated above, success in research science requires a dedication of time and effort. Taking time off to have a baby and working part-time while raising a family are not conducive to the fast track. Of course, one could argue that men could/should be equally involved in raising a family and that a woman should have the same opportunity to focus on career—and that is increasingly the case, at least in my experience. But for such a practical change to become the general de facto practice in research laboratories will require a rather dramatic change of mind-set. The current system almost always requires that women work harder than men to achieve the same goals. And unfortunately, in the face of this type of challenge, women often face more scrutiny (and are cut less slack) from their supervisors.

While this issue is a reflection of our society, there is another issue that may well be a function of (or at least heavily influenced by) biology. Women think and act differently from men. These differences, in turn, affect how a woman will do a particular job. For example, women (according to some studies) think more globally, while men think more linearly. Is one way better than another? Usually they are complementary—and this complementarity and diversity should be welcomed. But if a man is evaluating a woman's performance, he may have a problem in appreciating the contributions that have come from what to him seems to be a "strange" way of thinking. Women also tend to work better in groups than men, are more socially supportive, and work generally toward consensus. As currently structured, consensus is rarely the basis for decision-making (or data interpretation) in a research lab. Again, this difference in a woman's way of approaching a problem may not be understood by male colleagues or supervisors.

Because of the history of women in our society, there are general gender-specific features that may make life difficult for women. For example, women scientists—particularly as young trainees—are often more insecure than their male counterparts, have more difficulty in recognizing their own value, and tend to agonize more about criticism. As a result, they may wait longer to publish, be more reluctant to argue, and look toward reaching broad agreement rather than pushing their own agenda. In a field as competitive as research science, these traits can be problematic. Certainly, these descriptors don't apply to all women, and many men are also insecure and sensitive to criticism, but I would argue that these sex differences are real. The question is, then, whether the field can be sufficiently open to allow people (men and women) with all types of personalities, and diverse approaches to the world, to prosper and contribute.

This issue of diversity is, of course, much broader than simply considering male versus female. Indeed, while there is considerable progress toward integrating women fully into the research science field, there seems to be much less progress for individuals from some nonwhite ethnic backgrounds. Many minorities are significantly underrepresented in the laboratory—certainly as lab leaders, but also as research support personnel. I would argue that this problem is generally not one of active discrimination, although there are certainly problems with a tendency toward ethnic stereotyping. Rather, the problem stems from the poor educational (and self-image) preparation of individuals coming out of poorer communities. Training and preparation do make a difference when considering employment in science. And while there are now remedial programs to help selected individuals move forward more rapidly, the progress in integrating African Americans, Hispanics, and Native Americans within the research community has been very slow.

Should we be concerned about lack of opportunity for underrepresented minority trainees? And why does it matter? Aside from issues of general social responsibility, science is particularly dependent for progress on input from many sources, with different approaches and perspectives. If it is the case that individuals with different backgrounds do bring different perspectives to a problem, then diversity in science is not only "the right thing to do" but also critical for moving the discipline forward.

---

**Real-Life Problem**: You have just presented a poster at an international meeting. You are quite proud of your work and therefore devastated when a "colleague" from another institution comes by and comments that he's already done your experiments, using a newer and more powerful method. How do you deal with this input?

1. Try to find the weaknesses in your colleague's experimental approach.
2. Use his comments as motivation for generating new, more exciting results.
3. Discuss this "scoop" with your mentor, and try to devise novel experiments.
4. Reevaluate your place in your research community.

**Discussion**: Odds are that this type of experience will occur occasionally—even frequently—during a research career. While it's hard not to let it "get to you," it's not a basis for giving up on your research direction. It is certainly helpful to discuss this type of experience, as well as your reaction to it, with your mentor. He or she can help you see it in a more objective light, putting your work—and that of your new "competitor"—into a larger framework. You will inevitably want to learn more about what this colleague has actually done—whether he has actually "already done" your study, whether his techniques are indeed more powerful, and whether your results are similar. Rather than spend a lot of time searching for weaknesses, you might think about how your studies complement each other. Odds are good that your colleague's work will provide a basis for new and more interesting experiments—which you and your mentor can develop and execute, and which may provide even more exciting and significant results.

**Real-Life Problem**: As a young PI, you hire a postdoctoral fellow to work in your laboratory. You are near the same age, and the two of you become friends. Over the course of months, however, you become aware that he is not doing a very good job. What do you do?

1. Confront him, point out his shortcomings, and fire him.
2. Find another task for him to do.
3. Generate an excuse for letting him go ("lack of funding").
4. Provide him with additional training so that he can do a better job with the assigned task.

**Discussion**: Depending on the circumstances, all of the above possibilities may be useful alternatives. The fact that the PI and the postdoctoral fellow are friends obviously makes this situation more difficult, but it doesn't really change the nature of the problem or the tactics you might use to resolve it. While firing an individual who is not performing up to par is an intrinsic part of being a PI, such actions should almost always follow appropriate discussions with the individual, and a subsequent time period in which the individual has an opportunity to improve his or her performance. Firing an individual for poor performance usually requires consistent documentation of the individual's poor performance, as well as documentation of your discussions with the individual in which you've identified his or her shortcomings and provided a set of performance goals that you both agree upon. This discussion may involve a plan to provide the individual with further training that will enable him or her to meet performance goals. If the individual simply doesn't have the skills for the job he or she has been hired to do, it is sometimes possible to find another task. Flexibility to implement such an alternative may be restricted, however, if the

*Continued*

position is supported by a grant that requires the activities of someone with specific skills to do the assigned task. Many individuals will resign rather than risk being fired. And in some cases, a PI can find funding reasons for terminating the position (rather than firing the individual)—although again, if there are grant goals that require the activities of someone in that position, terminating the position can be tricky.

**Real-Life Problem**: You have begun your graduate dissertation work in Professor X's laboratory, and after several months your interactions with him have deteriorated significantly. How should you deal with this situation?

1. Confront your advisor and air your grievances.
2. Discuss the problem with the head of your graduate program, and see if you can change laboratories.
3. Avoid interactions with your advisor, and simply do your work to the best of your ability.
4. Work out a plan with another faculty member to move to her laboratory.

**Discussion**: The mentor–trainee relationship is an extremely important one. And as indicated above, dissatisfaction with one's adviser can actually arise from a variety of different issues. While it is always the ideal for advisor and student to like each other, that type of relationship isn't always necessary for the relationship to be productive. Further, these relationships tend to ripen with time; indeed, a fairly common phenomenon is for a student and advisor to become closer once the student has finished his or her dissertation work and "graduated" from the laboratory. At this stage, the mentor and trainee attain a more collegial relationship.

But for the student early in his or her laboratory experience, a less-than-friendly interaction can be a real problem. All of the possibilities listed above may provide useful strategies, but all have their potential problems. The least proactive of these alternatives—to keep your head down, stay out of trouble, and do your work—is worth trying, particularly if you're in a program that makes it difficult for you to move from laboratory to laboratory. Often, other interactions in the laboratory (e.g., with other students, with post-doctoral fellows) will in any case constitute most of the day-to-day contacts. Choice #3 is not a good option, however, if the result is likely to "turn off" your enjoyment of laboratory life. In that case, alternative approaches should be considered. A first step in that process might be a discussion with the graduate program advisor (choice #2). That discussion can provide the trainee with a sense of the program's philosophy regarding students moving from advisor to advisor. It can also result in some "leads"—that is, ideas about how to make the current situation more enjoyable/productive, or suggestions about other faculty members who might be willing to take in an unhappy camper.

I would urge students to be upfront with all these discussions. The current advisor may be insulted if he or she finds that a student is making plans to leave

*Continued*

the laboratory without first discussing this intention, and the initial advisor's unhappiness may be transmitted to other potential faculty advisors, who might subsequently become reluctant to take in the unhappy student. Thus, at some point, as painful as it might be, the student should have a frank discussion—not a "confrontation"—with his or her mentor. Ideally, that discussion should come early in the course of this growing dissatisfaction so that there's a reasonable chance that the relationship can be improved. If the issue causing distress revolves around the student's assigned tasks or dissertation direction, this discussion can go far toward resolving the problem and lead to a more enjoyable laboratory experience. If the problem arises truly from personality conflicts, then the discussion will serve simply to let the advisor know of the student's plans to find another laboratory home.

# 12 ◆

# Rewards and Riches

◆ What Kind of Rewards Do You Really Want From Your Job?

Before you jump into this (or any) profession, it's important to be clear about what you consider to be appropriate and satisfying rewards—and then see whether you are likely to attain those goals as a scientist. Choosing research science as a profession is not the way to become rich. And if you do become famous, it's likely to be a very restricted sort of fame, with your name familiar to a relatively small group of people. So what might you expect?

Certainly, your career rewards will depend on your personal priorities, the specific choices you make, and the kind of position you take. For example, there are specific positives and negatives to weigh when deciding whether to pursue a position in academics, in private industry, or in government. Typically, the salaries are better if you work for a for-profit company, but you may not enjoy the same level of intellectual freedom as in, say, academics. Job security may be greater in government than in industry (or academics), and financial support for your laboratory is likely to be less competitive. However, you may feel intellectually more isolated, restricted by laboratory security issues, and with less opportunity to interact with (and be challenged by) students.

There are, of course, exceptions to every generality of this type, and some of these differences across job sectors are becoming less marked. Further, it's possible—although sometimes rather difficult—to move from one type of environment to another if you find that you don't like a particular job type or framework. Despite these caveats, differences in job rewards do exist across employer types.

◆  Financial Compensation

I started this chapter by saying that you won't get rich doing research, but I should qualify that statement: research scientists can and do get rich—at least a few of them. In the past, these lucky few were those whose discoveries were parlayed into fantastically profitable products. But even these scientists rarely received the full benefit of their discoveries, since the patents for their discoveries were usually held by the institutions in which they were working (see the section on intellectual property below). As a result, most of the financial compensation flowed into the institutional (university, company) coffers. That scenario remains fairly common, although scientists are much more aware now of the potential reward for their work, and there is a growing tendency for researchers in many areas to patent virtually everything (from a genetically modified mouse to a sophisticated instrument) just in case some company comes along and is willing to pay them a lot of money for rights to their "invention."

Another, more direct way for research scientists to earn financial success is to parlay their expertise and imagination (and salesmanship) into the development of a "high-tech" company. Dozens of such companies have been started by scientists (with the help of venture capital), and they (and their colleagues) often reap significant profits. This money-making adventure is similar to starting other types of businesses but relies upon insights and future-thinking that are grounded on scientific knowledge and know-how.

So if your goal is to make money, there is a chance to do that as a scientist. However, few individuals choose a research career with that goal in mind: the chances are too slim. But if your financial goals are not so grandiose—if what you want is simply a comfortable living—being a research scientist has its attractions. First, salaries, while not extravagant (unless you get lucky and land a cushy job with a private company), certainly can provide more than a basic wage. And the salary is very likely to increase (sometimes dramatically) as your seniority increases. Most institutions that employ researchers—universities, private companies, the government—also offer attractive fringe benefits, including medical insurance and significant retirement programs. Some universities offer reduced tuition costs for the children of faculty. Many research positions are particularly stable, carrying a "tenured" label.

Perhaps most important, at least from my perspective, are the less tangible benefits of working in the laboratory. Some of them I've already mentioned—for instance, intellectual freedom (how much would you pay for that?), flexible scheduling, and stimulating colleagues. It's worth saying something also about the physical environment in which you work. Because laboratories are everywhere, in virtually every type of environment, the scientist does have considerable latitude in choosing where to live. Even the corporate research positions tend to have something of an academic feel, with an intellectual focus and people whose jobs involve developing and actualizing ideas. After all, research science is, more than anything else, an occupation of the mind.

♦ Intellectual Property

As indicated above, the financial rewards are generally modest in research science. However, the scientist does possess something extremely valuable, often referred to as "intellectual property." This property—knowledge that can be realized in (converted to) a potentially valuable commercial product—serves as the basis for a copyright or a patent. Converting intellectual property into a commercially marketable product has become a significant goal in many research institutions. Indeed, some institutions receive millions of dollars in income from patents that they've filed based on discoveries of their faculty. Intellectual property may not always be financially lucrative, but it can be a significant "reward" of research life. Because of its value, it is sometimes the basis of significant argument or even litigation. While this book is not the appropriate place for exploring the various forms and mechanisms to realize income from such property, it is worth discussing some of the common issues that arise in the laboratory with respect to intellectual property.

The most common forms of intellectual property are your experimental results. These data have value, since not only can they be published (to help establish your reputation) but they can also be used as a basis for further experiments, for preliminary data in grant applications, or even for patents. Most students are surprised to learn that, in most cases, they don't "own" the rights to their experimental data: those rights generally reside with the institution in which the student (or, indeed, the more senior investigator) has been working and/or with the granting agency that has provided the support for that work. When students finish their work and leave the lab, they cannot simply take their lab notebook or computer with them. That notebook, and the associated data, "belong" to the lab. Therefore, students (e.g., trainees finishing their Ph.D. work) and fellows should negotiate an agreement with the lab PI (who is the institution's "representative" in such matters) about what that PI is willing to share with the departing trainee. This process may not be insignificant, since data (i.e., intellectual property) may have considerable potential value to both the trainee and the PI.

When intellectual property has obvious financial value (e.g., the discovery of a new process for making "better" food crops, for generating energy more efficiently, for curing a disease), scientists often work with their institutions (e.g., the institution's "office of technology transfer") to facilitate the patent process and to help contact companies (or venture capitalists) who might have commercial interests in the discovery. The product/process can be patented by the institution, and it can in turn lease rights to the invention (according to a contractual agreement) to outside individuals or companies.

It is sometimes rather surprising to learn what kinds of discoveries can be patented and are thought to have commercial value. For example, not so many years ago, no one would have thought that a gene could be patented. But genes are being patent-protected regularly in today's world; by doing so, the discoverers of the gene guarantee that no one else can use that gene in their research (without the permission of the discoverer). Guidance about what is patentable, and what is likely to have commercial value, can generally be provided by experts in the institution's technology transfer office. This office is also the appropriate source of legal information that will outline the pathway for patenting your discoveries. Be advised, however, that just as indicated above for experimental data, discoveries that are made during the course of one's job in a research institution (yes, even a university) belong to that institution and the agency that funded the research. Granting agencies and research institutions currently negotiate agreements that cover the possibilities of sharing revenue made possible by a research discovery, and they are part of most grant application packets; that is, these agreements are in place long before any actual discovery is made. While the investigator who does the experimental work and makes the discovery is usually given some share of patent proceeds, the bulk of that income usually goes to the institution in which that discovery is made.

The major exceptions to these comments about income arising from intellectual property are the discoveries, or the experimental data, that are made/developed "privately"— that is, not in an institutional lab, not on company time, not using institutional equipment. In that case, the discoverer/inventor can obtain a personal patent and reap the financial rewards. You can see how this possibility can become legally confusing. For example, if an investigator works in a university setting for many years but then starts his or her own company to carry out research and make discoveries based on the "intellectual property" he or she accumulated at the university, to whom does that property belong? This type of question has been the basis of many lawsuits. And indeed, since it is no longer uncommon for an academic to start a private, for-profit company, that process usually involves some legal agreement with the university in which that scientist previously (or currently) was employed; such agreements identify what property can be "transferred" and how the profits will be shared once a marketable product is generated.

◆ Other Rewards

Beyond the financial and the intangible rewards mentioned above are a set of important factors associated with scientific research that should be explicitly identified (and appreciated). They include:

*Recognition*

Everyone likes to be acknowledged for a job well done. In the laboratory, individuals—whatever their level and whatever their task—are in most cases recognized for their contributions to a study. Indeed, in virtually every research report, there is a section called "Acknowledgements," where individuals who have contributed to the work—but are not authors on the paper—are acknowledged and thanked. In general, good-quality work is recognized by colleagues and coworkers, and even (perhaps especially) competitors. Since progress within the research hierarchy is primarily achieved via an apprenticeship-like system, such recognition is critical to the advancement of each individual. Your success is critical, too, to the reputation of the laboratory in which you're working, so everyone has a stake in seeing that credit is appropriately distributed. There are, of course, exceptions to this rather optimistic generality—situations in which the lab leader will take all the credit, or a colleague will downplay the contributions you've made. But for the most part, the system works pretty well.

With recognition comes prestige—for some, the greatest reward of all. To be sure, the recognition you earn will be from a rather narrow slice of the population, and therefore the prestige will be limited in scope. There are those who become sufficiently "famous" to gain wide notoriety; that occurs most frequently through the individual's efforts to popularize his or her work (or the work in the field) in a way that makes it appealing to the media and to a broad audience of nonscientists. There are relatively few of these "famous" commentators and experts; they tend to be the exception rather than the rule. Unfortunately, most of us don't seem to have the knack of (or perhaps the desire for) easy popularization. There is, of course, the fame achieved by Nobel Prize winners, who gain (at least momentarily) name familiarity across a fairly broad population. However, very few scientists are ever even nominated for a Nobel Prize, and even among the winners fame is generally fleeting—can you name even one Nobel Prize winner for science from last year? So fame, for most of us, comes in small doses, usually confined to specialists in our field—our coworkers, collaborators, and colleagues.

As in most fields, fame (that extreme form of recognition) usually is a result of not only real ability and accomplishment, but also a significant degree of luck and "public relations." As in other fields, some scientists are comfortable with the PR effort, but most are not. It is a useful skill and not one to be rejected out-of-hand. It is also a skill that can be learned (even among the shy and the tongue-tied). So whether or not your career plan includes attaining fame, learning to put yourself (and your field) out in front of the public should become a part of the job. As I've indicated in previous chapters, explaining scientific research in a way that is "user-friendly" is a sorely overlooked skill and a neglected opportunity. Whether your goal is to educate the public, to lobby for increased funding, or to change public policy, it will be important for you to develop your scientific theme in a way that is accessible to non-experts. And in the process, you never know—you might become famous!

### Security

Most of us are interested in the long-term stability of a job in a particular field. Job security for research scientists is, by most standards, one of the great attractions of the profession. Why should that be so? Well, to start with, it is unlikely that we'll ever answer all the questions that need to be addressed, that we'll ever reach a societal state in which everyone agrees that we've come far enough (and that, therefore, there's no longer any need for research). And even if interest in a field-specific subject wanes, there will always (and I say this hopefully) be a need for individuals who can think like scientists. When that is no longer the case, then all bets are off.

In terms of specific job positions, security varies somewhat depending on your employer, on your seniority, and on the source of your funding. Grant funding goes through cycles, and in times of plenty, job security never seems to be a problem. During times of restricted funding, however, even the concept of "tenure" may not be protective. For example, a senior tenured professor at a university may be able to continue earning a salary when the grant funding dries up, but he or she won't be able to run a laboratory. And in times of want, competition for research funding (which is almost always reasonably tough) becomes fierce. Downsizing in industry, or even in government, may pull the job rug right out from under you.

These words are not meant to scare you away. They are meant as a warning—to stop and consider what you actually need in terms of security, and to consider whether your needs are realistic. If you consider the alternatives in other professions, job security for research scientists is unusually supportive. A researcher in private industry may have salary and security guarantees that you won't find in academics—or there may be no safety net at all. Both industry and academics offer an expanding job market, but you have to be sufficiently clever to find your correct fit and to sell yourself as indispensable. As our societies grow and change, there will always be a need.

*Friendship*

Personal interactions with your colleagues can be among the most rewarding aspects of the research business. As a student, you will develop valuable friendships with your fellow trainees—friendships that can last a lifetime. These relationships, and those with your mentor, are critical not only for the key roles that these people will play in your training, but also for their contributions to your subsequent career. If you are lucky, mentors and fellow students will become colleagues and collaborators. They will support you personally and professionally. Similarly, as you progress to more senior positions, those interactions with colleagues and coworkers will provide you with intellectual stimulation, sympathetic (and ideally insightful) review of your work, and an invaluable framework that supports your personal goals and ambitions. While they may be competitors, they are (again, if you're lucky) also often your biggest supporters.

The establishment of this extended family offers a number of nonscientific rewards. For example, having friends and colleagues from all parts of the world may offer wonderful opportunities to travel and explore. For a young scientist, attending a conference in a far-off country is a wonderful reward. To attend such a meeting and have your foreign friend serve as a local guide makes this experience truly special. You will find it easy to establish close personal bonds that start with your common intellectual commitments.

*Fulfillment and growth*

It's worth pausing here to consider also the value and reward—albeit not monetary—in learning to do something well. This type of reward—self-respect and satisfaction—is obviously not peculiar to scientific research, but research offers many opportunities for achieving that type of reinforcement. There are technical skills to hone, intellectual facility to aspire to, people skills and plain old management skills to develop. The research profession is for those who thrive on challenges—technical, intellectual, and personal. Your world in the laboratory can be as broad as you want it to be—and the rewards are thus equally varied.

One of the most cherished (at least for me) perks of research is the opportunity to be your own boss. That is certainly the case for those who choose the academic path but is also a significant aspect of research activities associated with other environments. Not only is your time up to you to schedule, but so too can you determine the nitty-gritty of your job. To a greater or lesser extent, the focus of your research is yours to choose—and if you can sell it to those who pay your way, it is yours to develop.

To be successful in the research environment requires flexibility, a willingness to make the best of your opportunities. For those who take advantage of these opportunities, the experience gives rise to growth—not only as a scientist but also as an individual. Just as in other walks of life, research topics come in and out of fashion, sometimes being "hot" and other times being viewed as "old-fashioned." The "in" topics are determined by the challenges of the day, as well as by the new technologies that make timely the questions that were previously unapproachable. You must grow as your chosen field expands—and in turn you have the opportunity to contribute to further growth and understanding of your special topic. This give-and-take process can be truly exciting. The fulfillment one experiences in making substantive contributions to our knowledge and to the development of our society—be it in medical practice, energy or food production, or some other more esoteric topic—is for many individuals worth the work and frustration that has inevitably accompanied the journey. If you are one of those who finds joy and satisfaction in learning, growing, and contributing, then research science can be a wonderful way to spend your life.

**Real-Life Problem**: As a young scientist, you have been offered several research scientist positions—at a large university, in a pharmaceutical company, and in a government laboratory. How do you decide what job to take?

1. Ask your mentor and follow his or her advice.
2. Take the job that offers the best salary and benefits package.
3. Take the job that gives you the best laboratory support.
4. Weigh the relative rewards in each environment as they relate to your own goals.

**Discussion**: This is a nice problem to have—and not everyone has the luxury of struggling with such a decision. As a young scientist, it's always helpful to seek the perspective of a more senior individual, especially an individual whose ideals and goals appear to be similar to your own. A senior advisor can help you "interpret" the offers, read between the lines, and identify the pros and cons of each offer. Clearly, however, you are the one in the best position to decide what is "right" for you. Each offer should be evaluated against your personal goals and preferred lifestyle. For some, the salary and benefits may be an overriding issue. For others, laboratory independence may be key. For still others, the opportunity to work with a specific set of colleagues and students, or to focus on a particular problem, may be the determining factors. By the time you get to this stage of your career, you should have a pretty good idea of what you want. Remember, whatever your choice, you're likely to be doing it for a long time, so choose something that you'll enjoy.

**Real-Life Problem**: Your work in the laboratory has resulted in what you believe to be a valuable, patentable discovery. You think, in fact, that you may be able to realize a significant financial gain, but have been cautioned not to be too optimistic about this potential. Why should you not expect to make a fortune from your discovery/invention?

1. There is no commercial interest in your discovery.
2. The institution within which you've done the relevant work will be the major patent holder.
3. There are many other individuals trying to patent a similar product, and the legal issues are complex.
4. Your discovery is not patentable.

**Discussion**: Not every discovery is patentable. Sometimes that's because others have already made a substantially similar discovery. Sometimes it's because the "discovery" is too general and does not translate into a clearly definable piece of intellectual property. Sometimes it's because it involves materials and processes that are already in the public domain. But more

*Continued*

and more, investigators are finding that their institutions are more than happy to help them patent an
interesting discovery, especially if it has potential
financial implications. If it does not (and the institution will make that assessment), the institution may
not want to go through the effort and expense of such
a process. It is often up to you, the discoverer, to make
the case for the potential of financial returns. Your
institution's administrators may need to be convinced
that your discovery is new (they will not want to
become involved in legal fights as to rights) and has
significant commercial potential. And if you can convince your institution to pursue a patent, remember
that the institution (and granting agency) will reap
the primary financial benefits, not you.

# Concluding Thoughts

As I indicated at the beginning of my discussions, the impetus behind this book was to provide the young scientist with some of the information that could help guide his or her decision about whether to pursue this career—and how to negotiate some of its challenges. It is a personal view and therefore reflects my personal biases. Despite these biases, I have made an effort to be relatively even-handed with respect to the question of whether one *should* enter a career in research. I've offered both upsides and downsides, but now it is time to "come clean."

We, as scientists, currently find ourselves in a unique but confusing position. On the one hand, technological advances have made previously only-imagined discoveries and insights almost everyday occurrences. There is tremendous excitement about the *potential* for meaningful advances. On the other hand, social pressures have often relegated scientists to second-rate positions over the past couple of decades. While we scientists are in part responsible for the development of that attitude, much of it has come from our society's political and social leadership (or lack thereof). Earlier in my career, my mantra to students and fellows was very positive—if you work hard and do good work, you will be rewarded and be successful. It has been difficult for me to maintain that position in recent years.

I feel that tide is now once again turning, and that there is a great wave of support coming our way. Politicians now talk about the need for key social decisions to be informed by, and based on, real scientific data. If one believes the talk (and I do), then there will be a significant change in the role that science plays in our society. Undoubtedly, any such change will come gradually. But if change is realized, the opportunities for those in research science careers will be immense. It could be an extraordinarily exciting and rewarding time to be a scientist.

I am optimistic, and I hope you will be, too.

# INDEX ◆

Principal investigator (PI) (*Con'd*)
    as career choice, 26–28, 32
    as "face" of laboratory, 25, 29
    leadership, 25, 117–18
    responsibilities, 25, 26–27, 156
    skills, 27
    traits, 27–28, 31
Probability and statistics, 34, 36,
        43–44, 144, 151
    *See also* Prediction
Publishing scientific papers
    authorship, 63–64, 70–71
    changes in scientific publication,
        66–68
    ethical conduct, 68, 125, 132–33
    factors affecting where to
        publish, 60–62
    by graduate students, 10
    impact factor, 61–62
    manuscript reviewing, 64–66,
        67, 68
    open access movement, 67–68
    personal online publication, 67
    target readership, 61
    *See also* Journals; Scientific papers;
        Writing scientific papers

Reputation
    conservative and incisive
        thinking, 39, 45, 65
    of graduate advisor, 9
    of graduate programs, 7
    importance for successful
        career, 22, 45
    and manuscript reviewing, 65
    of postgraduate advisor, 13
    presentations and talks, 79, 80
    prestige, 108–9, 175
    risks from problems with
        publishing, 70, 71
    of scientific journals, 61–62
Research science as career choice,
        overview, 28–29, 183–84
    *See also* Rewards of
        science careers

Responsible conduct of research
        (RCR), 123–24, 128, 130
    *See also* Ethical conduct
Resumes, 16
Reviews, as type of scientific
        writing, 52–53
    *See also* Peer review; Scientific
        papers
Rewards of science careers
    choosing type of employer,
        169–70, 180–81
    development of high-tech
        companies, 171, 174
    fame, 175–76
    financial compensation, 169,
        170–72, 181–82
    friendship, 154, 155–56,
        164–66, 178
    fulfillment and growth, 178–79
    intangible benefits, 171–72
    intellectual property, 129–30,
        172–74, 181
    Nobel Prize, 108, 176
    overview, 169–70
    patents, 129, 170, 172–74, 181–82
    personal rewards of creative
        enterprise, 139–40
    prestige, 108–9, 175
    primacy, 126, 129–30
    recognition, 175–76
    security, 170, 176–78
    tenure, 171, 177
    *See also* Career choices
Role of scientists in society
    communication with public,
        146–48, 150, 151–52
    engagement of scientists in
        society, 130–32, 133–34
    future of science in
        society, 150–51
    historical basis for scientific
        creation, 136–37, 143–44
    overview, 143–44, 150–52, 183–84
    public distrust of research science,
        123, 145–46, 149–50

relationship with history of issues, 45, 50, 55, 143

responsibility of scientists to society, 132, 133–34, 145, 146–48

*See also* Ethical conduct

Science, definitions and functions, 4, 144–46

*See also* Scientific thinking

Scientific method, 34, 145, 148, 152

Scientific papers

abstracts, 50

acknowledgements section, 52, 175

discussion section, 51–52

figures, 56–57

by graduate students, 10

introduction section, 50–51

letters to the editor, 53

methods section, 51

overview, 49–50

parts of a paper, 50–52

results section, 51

reviews, 52–53

supplemental information posted online, 52

tables, 56–57

*See also* Publishing scientific papers; Writing scientific papers

Scientific research as a creative enterprise. *See* Creative enterprise

Scientific thinking

as acquired art, 33–34, 47, 48

challenges, 38–39

as creative process, 37–38

critical thinking, 46–47

data collection, 35–36

difference between correlation and causation, 36–37, 46, 148

experimental design, 34, 37–38, 40–43

hypothesis development, 34, 37–38, 40–41, 90, 91–92

impossibility of proving truth, 34, 41, 47–48, 131, 147

overview, 33–38, 46–48

scientific logic, 34

scientific method, 34, 145, 148, 152

sufficient and necessary evidence to support hypothesis, 34

*See also* Interpretation of data

Self esteem/self-confidence, 156–58, 163–64

Self-promotion, 16, 80–81, 157–58

Social aspects of scientific research. *See* Politics of science; Role of scientists in society

Squatting, 120–21

Talks. *See* Presentations and talks

Tenure, 171, 177

Truth, impossibility of proving scientifically, 34, 41, 47–48, 131, 147

Vaccinations and autism, 36–37

Verbal presentations. *See* Presentations and talks

Websites, 67, 150

"Why" questions, 4–5, 107, 145, 148

Women in research science, 160–62

Writer's block, 53–54, 58

Writing scientific papers

avoiding assumptions about readers, 56

drafts, revisions, and corrections, 55, 56, 57–58, 69

easy-to-understand constructions, 59

effective language and grammar, 59

familiarity with literature and work in field, 59–60

figures, 56–57

general aspects of effective writing, 58–60

history of problem, 45, 50, 55

Writing scientific papers (*Cont'd*)
how to start writing,
53–56, 69–70
linking ideas, 59–60
organization, 58, 60
outline method, 54, 58
presentation of data, 56–57

stream of consciousness
approach, 54–55, 58
tables, 56–57
telling a story, 45–46, 55, 69, 76
tips about writing, 57–60
*See also* Publishing scientific
papers; Scientific papers